A Monte Carlo Primer
Volume 2

A Monte Carlo Primer
Volume 2

Stephen A. Dupree

and

Stanley K. Fraley

Sandia National Laboratories
Albuquerque, New Mexico

Kluwer Academic / Plenum Publishers
New York, Boston, Dordrecht, London, Moscow

Additional material to this book can be downloaded from http://extras.springer.com

Library of Congress Control Number: 2002277146

ISBN 978-0-306-48503-9

233 Spring Street, New York, New York 10013

http://www.kluweronline.com

10 9 8 7 6 5 4 3 2 1

A C.I.P. record for this book is available from the Library of Congress

To Pat and Huri

Preface

In Volume 1, *A Monte Carlo Primer – A Practical Approach to Radiation Transport* (the "*Primer*"), we attempt to provide a simple, convenient, and step-by-step approach to the development, basic understanding, and use of Monte Carlo methods in radiation transport. Using the PC, the *Primer* begins by developing basic Monte Carlo codes to solve simple transport problems, then introduces a teaching tool, the Probabilistic Framework Code (PFC), as a standard platform for assembling, testing, and executing the various Monte Carlo techniques that are presented. This second volume attempts to continue this approach by using both custom Monte Carlo codes and PFC to apply the concepts explained in the *Primer* to obtain solutions to the exercises given at the end of each chapter in the *Primer*.

A relatively modest number of exercises is included in the *Primer*. Some ambiguity is left in the statement of many of the exercises because the intent is not to have the user write a particular, uniquely correct piece of coding that produces a specific number as a result, but rather to encourage the user to think about the problems and develop further the concepts explained in the text. Because in most cases there is more than one way to solve a Monte Carlo transport problem, we believe that working with the concepts illustrated by the exercises is more important than obtaining any one particular solution. Of course, having said this, it is nonetheless important to be able ultimately to obtain particular, and correct, solutions to real-world problems using Monte Carlo transport methods.

The objective of this book is to indicate several approaches that can be used for solving each of the exercises, to use the process of obtaining such solutions to illustrate the techniques presented in the *Primer*, and to offer insights into how one may be confident that one has obtained an acceptable answer. While a guarantee of accuracy is possible only in simple situations, confidence in the answer one obtains is always important and the more thoroughly one understands a problem the more confidence one can gain that the correct solution has been obtained.

Because frequent reference is made in this book to the exercise numbers, tables, figures, and equations given in the *Primer*, the following conventions are used:

> Examples: All examples cited herein refer to the example problems described and solved in the *Primer*. The numbering of the examples is identical to that used in the *Primer*.
>
> Exercises: All exercises are cited by two numbers separated by a period. The first number indicates the chapter in the *Primer* from which the exercise is taken, and the second indicates the exercise number for that chapter. Thus Exercise 2.3 designates Exercise 3 of Chapter 2 of the *Primer*.
>
> Tables, figures, and equations: All references herein to the tables, figures, and equations that appear in this book are specified using two numbers separated by a period. Thus Table 2.5 refers to the fifth table of Chapter 2 of this book. All references herein to sections, tables, figures, and equations that appear in the *Primer* are cited in the same manner except they are followed by the letter "P." Thus Table 2.5P refers to the fifth table of Chapter 2 of the *Primer*.

This book assumes the reader has a knowledge of integral calculus, neutron transport theory, and Fortran programming. It would also be useful for the reader to have available a PC with a Fortran compiler. All of the Fortran routines presented reflect the authors' styles and contain constructs that include both Fortran 90 and older versions of Fortran coding. Hopefully readers will be able to follow these constructs easily and, if desired, change them into a form compatible with their own programming style. When referred to in the text, Fortran subroutines and variable names are designated by single quotation marks. Exponents of computed results may be designated in the text either by using a "+" or "-" sign with the base 10 understood, either with or without the designators "D" or "E;" or by using the base 10 with a superscript. Thus the expressions 1.0+5, 1.0E+5, and 1.0×10^5 all designate the same quantity. Because compilers and execution speeds vary, the reader should not expect to reproduce the run times cited, although the relative run times obtained in the various calculations should be similar to the present results.

For the convenience of the user, a compact disk is included with this book. This disk contains copies in ASCI format of all of the Fortran routines presented in tabular form both in the *Primer* and in the current volume. For a description of the files on the CD see the 'readme.txt' file on that disk.

We would like to thank the editor of the previous volume, Ingrid Gnerlich, for suggesting the current volume. We hope the users of the *Primer* will find this work useful. We would also like to thank our colleagues Patrick McDaniel, Thomas Laub, and Leonard Connell, for reviewing this volume. Their comments and suggestions have greatly improved the accuracy and completeness of the solutions and discussions as well as the clarity of the text. Any errors remaining are, of course, the responsibility of the authors.

Stephen A. Dupree
Stanley K. Fraley
Albuquerque, NM
November 2003

Contents

Chapter 1

Introduction

Exercise 1.1

Statement of the problem

1. The probability of throwing "box cars" (two sixes) with a standard pair of dice is $1/6^2 \approx 0.02778$. Use Monte Carlo sampling to verify this result and to determine the probability of throwing box cars twice in a row.

Solution

To solve this exercise using Monte Carlo sampling methods we must model the throwing of a pair of dice in order to obtain a mathematical realization of a possible series of throws. By examining the results of many such throws we can then determine the probability of throwing box cars. To develop the model we build on the procedure for predicting the average outcome of a physical process that was discussed in Example 1.1.[a] As described in that example we assume there is an unlimited supply of random numbers available in the range (0,1) and we use these random numbers to predict the outcome of a hypothetical, or mathematical, throw of a pair of dice. As in Example 1.2 we use the function 'fltrn,' described in the Appendix of the *Primer*, to obtain our random numbers. As an aside, although the problem specifies that we use two values of six on the dice, or

[a] As explained in the preface, citations such as this refer to Volume 1, *A Monte Carlo Primer, A Practical Approach to Radiation Transport*, Kluwer Academic/Plenum Publishers, New York, 2002, which hereafter is referred to as the "*Primer*."

box cars, as the tally for our calculation, the probability of throwing two sixes is equal to the probability of throwing any other pair of numbers on the dice, so the result is valid for any such pair of numbers, not just for two sixes.

We can represent the numbers one through six, which we assume correspond to the faces of a standard die, by defining equal sub-intervals over the range (0,1) of an independent variable. In this case a result of "one" can be specified by the interval (0,1/6], a result of "two" by the interval (1/6,2/6], etc. Thus a "six" corresponds to the interval (5/6,1). Because a random number selected over the range (0,1) will necessarily fall within one of these ranges, each such random number will uniquely identify one face of a die. If our random number generator is of good quality these faces will occur with equal probability and in random order.

Given the above mathematical representation for the faces of a die based on random numbers selected over the interval (0,1), we must convert the values selected from within this range to an appropriate set of integers over the range one through six. To do this we multiply the random numbers obtained within the interval (0,1) by six, use the integer portion of the result, and add one. This process produces a random integer over the desired range and thus uniquely identifies a score for the throw of a die. When our integer is equal to six we assume that our throw has produced a value of "six." That is, for each throw of a die we select a random number ξ, and if $[[6\xi]]+1 = 6$, where $[[x]]$ represents the integer portion of x, we say the throw has produced a "six." Because dice are normally thrown as a pair rather than one at a time, to get box cars we assume one must throw two sixes in a row, and they must fall as a pair. That is, we cannot just wait for a result of six and then see if the next throw is also a six since the first six may have represented the second die of a pair.

To obtain sequential throws of box cars we must have four consecutive random numbers greater than 5/6 that correspond to two throws of a pair of dice. Because box cars constitute a single event, it is correct to wait until one throw of box cars occurs and then see whether the next throw of two dice produces a second result of box cars.

The Fortran code shown in Table 1.1 provides a means for modeling a series of throws of a pair of dice, of estimating the probability of throwing box cars, and of estimating the probability of throwing box cars twice in a row. The standard deviation in the results obtained from this code is calculated using eqn 1.9P in the same manner used in Examples 1.2 and 1.3.[b] In the coding of Table 1.1 we use the default seed for the random

[b] Although the formula for the standard deviation of a binomial distribution could have been used instead of eqn 1.9P, the latter is not introduced in the *Primer* until Chapter 3, Table 3.8P. A "P" included in a citation in this book refers to the indicated item in the *Primer*.

number generator 'fltrn' rather than allowing the user to enter a seed. The latter is a desirable refinement when diagnosing or evaluating complex calculations but will be forgone for the present. The user can add the feature if desired (see various examples in the *Primer*).

Running the program of Table 1.1 for a sampling of 10^8 throws gives a probability for box cars of 0.0277946 ± 0.0000164. For throwing box cars twice in a row we obtain a probability of $(7.7122 \pm 0.0278) \times 10^{-4}$. It is easy to solve this problem analytically. The correct answers are, for single box cars, $p_1 = 1/6^2 \approx 0.027778$, and for double box cars, $p_2 = (1/6^2)^2 \approx 7.716 \times 10^{-4}$. Both of the calculated results are within one standard deviation of these exact answers.

Table 1.1. Fortran Program for Exercise 1.1

```
DOUBLE PRECISION fltrn
icountbc=0;  icountdbl=0;  isumsqbc=0;  isumsqdbl=0;  n=100000000
DO i=1,n
  n1=INT(6.0*fltrn())+1
  n2=INT(6.0*fltrn())+1
  IF(n1.EQ.6.AND.n2.EQ.6) THEN
    icountbc=icountbc+1;  isumsqbc=isumsqbc+1
    n3=INT(6.0*fltrn())+1
    n4=INT(6.0*fltrn())+1
    IF(n3.EQ.6.AND.n4.EQ.6) THEN
      icountdbl=icountdbl+1;  isumsqdbl=isumsqdbl+1
    ENDIF
  ENDIF
END DO
stdev1=SQRT((isumsqbc/FLOAT(n)-(FLOAT(icountbc)/FLOAT(n))**2)/FLOAT(n))
stdev2=SQRT((isumsqdbl/FLOAT(n)-(FLOAT(icountdbl)/FLOAT(n))**2)/FLOAT(n))
WRITE(*,*)icountbc,icountdbl
WRITE(*,*)FLOAT(icountbc)/FLOAT(n),' + or -',stdev1,' versus',1./36.
WRITE(*,*)FLOAT(icountdbl)/FLOAT(n),' + or -',stdev2,' versus', 1./(36.*36.)
STOP
END
```

Discussion

The results obtained for this exercise depend on how the reader interprets the statement of the problem. It may be appropriate to consider the changes that would result in the coding and in the answers if the dice were thrown individually or if they were thrown four at a time. In the latter case one would need to calculate the probability that all four dice thrown simultaneously would be sixes, or the chance that at least two of them would be sixes. One could also consider the change that would occur if the result sought is for two and only two box cars in a row. In the latter case one can show that $p_2 = (1/6^2)^2 - (1/6^2)^3$.

The representation of the values one through six by the intervals (0,1/6], (1/6,2/6], etc., as used in this exercise, is not unique. We could as easily have assigned a value of "six" to an interval between any two numbers a and b, $0 \leq a < b \leq 1$, for which $b - a = 1/6$. We could also have segmented the interval to which we assign the value "six" rather than defining a continuous interval, provided the sum of the lengths of all such segments still equals 1/6. However, using the continuous interval (5/6,1) to represent the value "six" is as good a choice as any and makes the coding simple.

Exercise 1.2

Statement of the problem

2. Using the rejection technique, calculate the area under a half wavelength of a sine wave,

$$y = \int_0^{\pi} \sin x dx$$

Estimate the standard deviation using eqn 1.9P, and show that the standard deviation is reduced by approximately a factor of two if the number of samples taken is increased by a factor of four.

Solution

This problem is closely related to Example 1.2. In that example we select two values, x and y, at random over the interval (0,1) and compare the square of the radius of the resulting sample point (x,y) in two dimensions, $r^2 = x^2 + y^2$, to the unit radius 1.0. If $r^2 \le 1$ we accept the sample point and add a value of one to a tally variable; otherwise we reject the sample point. That is, if the random point (x,y) in the unit square $0 < x < 1$ and $0 < y < 1$ is on or below the curve $x^2 + y^2 = 1$ we accept the sample point, otherwise we reject the sample point. We then select a new pair of random numbers and continue the calculation. The ratio of the number of samples accepted to the total number of samples taken provides an estimate of the area of one quarter of a unit circle.

To use the rejection technique in the current exercise in a manner analogous to that of Example 1.2 we must select a sample point at random from within the rectangle defined by the upper and lower bounds of the independent variable x and the range of the integrand over these values of x. The limits of x are defined by the bounds of the integral, $0 < x < \pi$. Over this range of x the integrand can take on values between zero and one. Thus if we define our rectangle in a two-dimensional space (x,y) we wish to select points randomly over the range $x \in (0,\pi)$ and $y \in (0,1)$. To allow x to vary over the range $(0,\pi)$ instead of (0,1) we use the technique described in the *Primer* on p. 16, following eqn 1.13P. That is, the selection of points is typically accomplished by multiplying random numbers picked over the interval (0,1) by the actual width of the interval and adding the lower bound. This is similar to what was done in Exercise 1.1 in order to select an integer within the range one to six. The explicit mathematical expression for this transformation is shown in Chapter 2 of the *Primer* as eqn 2.20P.

To reiterate, in order to select a random number ξ^* over the interval (0,a), where $a \neq 1$, we select a random number ξ over the interval (0,1) and calculate

$$\xi^* = a\xi \tag{1.1}$$

To solve the current exercise using the rejection method we first select a point (x,y) within the rectangle $x \in (0,\pi)$ and $y \in (0,1)$. To obtain a random number ξ^* within the interval $(0,\pi)$ we use eqn 1.1 with $a = \pi$. Having obtained the point (x,y) we compare sin x to y. If $y < \sin x$ we add one to a tally variable; otherwise we do not. We then select a new pair of values for x and y and continue the calculation. The ratio of the number of points that are accepted to the total number chosen, times the area of the rectangle from which these points are chosen, is an estimate of the value of the integral.

A Fortran code to execute this problem is shown in Table 1.2. In using eqn 1.9P to determine the standard deviation of our result we take advantage of the fact that, since the score to be added to the tally is always unity, the average of the sum of the squares of the scores is the same as the average of the sum of the scores. Running the code of Table 1.2 for 10^6 sample points produces the result 2.001125 ± 0.001511. Using 4×10^6 sample points produces the result 1.999976 ± 0.000756. Thus increasing the number of samples by a factor of four reduces the standard deviation by almost exactly a factor of two.

Discussion

Since eqn 1.9P has a factor of $1/\sqrt{n}$ it is to be expected that the estimate of the standard deviation will be reduced by a factor of two if the number of samples taken is increased by a factor of four. However, this is only true if sufficient samples have been taken to obtain a valid estimate of the standard deviation, and if the standard deviation is finite. A case in which the standard deviation is not finite is addressed in Chapter 7 of the *Primer*.

If we define the "efficiency" of a Monte Carlo calculation to be a measure of the number of computer manipulations required to obtain a result of a given accuracy, then it is clear that the rejection technique, as used here, is not a very efficient means for solving this problem. This is because the probability of selecting a point under the sine curve over the interval from 0 to π is equal to $2/\pi$. As the fraction of points that are rejected increases, the efficiency of a rejection calculation decreases. Therefore the rejection technique is usually reserved for problems in which a formulation can be derived for which the probability of rejecting a point is small, or for problems in which other methods of solution are difficult to employ. The method used to solve Example 1.3, in which the value of an integral is estimated by calculating the mean value of the integrand over the range of the independent variable and then multiplying this mean by the range of the independent variable, would be more efficient than the rejection technique for solving the present exercise. In the method of Example 1.3 every evaluation of the integrand is used in the estimate of the value of the integral rather just those that pass a rejection test.

Table 1.2. Fortran Program for Exercise 1.2

```
      DOUBLE PRECISION fltrn
      pi=3.14159265
      WRITE(*,*)
    2 WRITE(*,'(1X,A\)')' Enter number of samples '
      READ(*,*)NMAX
      IF (NMAX.EQ.0) STOP
      nscore = 0
      DO 1 i = 1,nmax
        x = fltrn()*pi
        y=fltrn()
        if(y.LE.SIN(x))nscore = nscore+1
    1 CONTINUE
      prob = FLOAT(nscore)/FLOAT(nmax)
      avgsumsq=prob
      sig = SQRT((avgsumsq-prob*prob)/FLOAT(nmax))
      area=pi
      answer=prob*area
      stdev=sig*area
      WRITE (*,'(1X,A\)') ' Answer and StDev '
      WRITE(*,*)answer,stdev
      GO TO 2
      END
```

In our solution of Exercise 1.2 we deliberately choose the limits of our range for y to be equal to the range of the integrand over the limits of the independent variable. This makes the size of the rectangle from which we choose our sample points as small as possible. We are not constrained to this range, however. If we wish we could choose to select y from the range (0,2) instead of (0,1). In such a case, by including values of $y > 1$, which we know cannot be less than sin x, we would decrease the efficiency of our calculation. Nevertheless, we would still be able to estimate the value of the definite integral correctly because we would multiply the new area of our rectangle, which is larger than the one we actually used, by the probability of a point selected from that larger rectangle lying under the same integrand. This new probability would be smaller than that previously calculated by the exact amount required to correct for the new area of the rectangle. It is for this reason that we must normalize our result by multiplying the ratio of the probability of a point being accepted by the area within which the points are selected. Such normalization allows us to maintain a "fair game" of Monte Carlo, a concept that is defined in Chapter 2 of the *Primer*.

It is not necessary to use a rectangular domain to apply the rejection technique. By employing a transformation of variables a wide range of options for applying this technique can be introduced. Generally such transformations require additional computer manipulations, which will affect the efficiency of the calculation. A practical definition of the efficiency of Monte Carlo calculations is introduced in Chapter 6 of the *Primer*.

Exercise 1.3

Statement of the problem

3. The normal curve is given by

$$\Phi(x) = \frac{1}{\sqrt{2\pi}} e^{-x^2/2}$$

Evaluate

$$I = \int_{0.25}^{4} \Phi(x)dx$$

using Monte Carlo and show that $I \approx 0.4$. Estimate the standard deviation of your result.

Solution

To solve this problem we use the method of Example 1.3. However, in the present case we need to select numbers randomly over the interval (0.25,4). Hence it is necessary to derive a means of selecting points randomly over an interval that is not only greater than one, and thus is wider than that covered by standard random number generators, but also does not begin at the origin. To accomplish this, we could use eqn 1.1 with $a = 4$ and reject any points with values less than 0.25. Although this would be correct, a more efficient method, in the sense defined above, would be to transform the interval (0,1) into the interval (0.25,4) directly. To do this we must not only use eqn 1.1 to transform the points ξ into an interval of the correct width by using a = 4 - 0.25, but we must also translate the resulting points ξ^* to the correct coordinates by adding the lower limit of the interval, 0.25. That is, the most efficient way to select points from the interval (a,b) is to modify eqn 1.1 to the form

$$\xi^* = a + (b-a)\xi \tag{1.2}$$

This revised version of eqn 1.1 is given by eqn 2.20P.

Using the transformation given by eqn 1.2, or an equivalent formulation, the solution to this exercise follows from eqns 1.10P and 1.11P. The standard deviation of the Monte Carlo result can be estimated from eqn 1.9P, as shown in Table 1.5P. A Fortran code to estimate both the value of the integral I and the standard deviation of the result is given in Table 1.3. Running 10^8 sample points using this code gives a result for I of 0.401207 ± 0.000047. We surmise that the actual value of I is close to but slightly greater than 0.4.

Table 1.3. Fortran Program for Exercise 1.3

```
      DOUBLE PRECISION fltrn,sum,sumsq,score
      pi=3.14159265
      WRITE(*,*)
    2 WRITE(*,'(1X,A\)')' Enter number of samples '
      READ(*,*)NMAX
      IF (NMAX.EQ.0) STOP
      sum=0.0d0;  sumsq=0.0d0
      DO 1 i = 1,nmax
         x=0.25 +3.75*fltrn()
         score=3.75*EXP(-x*x/2.0)/SQRT(2.0*pi)      !  3.75 is Delta x to normalize
         sum=sum+score
         sumsq=sumsq+score**2
    1 CONTINUE
      sum=sum/FLOAT(nmax)
      sumsq=sumsq/FLOAT(nmax)
      stdev=SQRT((sumsq-sum**2)/FLOAT(nmax))
      WRITE(*,*)' Integral= ',sum,' sigma=',stdev
      GOTO 2
      END
```

Discussion

One could use the rejection technique to solve this problem. It is easy to show that the maximum value of $\Phi(x)$ for $x \in (0.25,4)$ is $\Phi(0.25) \approx 0.154258....$ Thus as in Exercise 1.2 one could select a random point (x,y) within the rectangle $0.25 < x < 4$ and $0 < y < \Phi(0.25)$ by selecting two random numbers (x,y) within these ranges. One would then test whether $y < \Phi(x)$. If so a score would be recorded by adding a value of unity to a tally. The user would then select another point (x,y) and repeat the test. The value of the integral I would be equal to the probability of the points selected lying below the curve $\Phi(x)$ (i.e., the value of the tally divided by the number of points tested) times the area of the rectangle formed by the limits of the ranges of x and y.

As discussed under the solution to Exercise 1.2, the upper limit for y need not be exactly equal to $\Phi(0.25)$. Instead, for convenience, we can add a small amount to our estimate of $\Phi(0.25)$ and be assured that our rejection calculation is normalized to the correct result. For example, having determined $\Phi(0.25) \approx 0.154258...$, we can select the slightly larger value of 0.16, say, for the upper limit of y. In this way we can guarantee a correct solution without significantly decreasing the efficiency of our calculation, as defined above, and we need not undertake the effort to calculate the minimum upper limit of the range of y to a large number of significant digits.

Although the rejection technique would provide a valid solution to this exercise, if it were used on a rectangular domain it would be less efficient than the method of Table 1.3. Just as in Exercise 1.2, the integrand would have to be evaluated at each point x that is selected in the calculation, but not every such point would be used in the estimate of the result. Instead, any point for which $y(x) > \Phi(x)$ would be rejected. Therefore computer time would be spent evaluating quantities that do not contribute directly to the final result. The fraction of all points tested that are not rejected will depend on the fraction of the area in the rectangle $0.25 < x < 4$ and $0 < y < \Phi(0.25)$ that is under the curve $\Phi(x)$. In the present case this is approximately $0.4/0.58 \approx 0.69$; i.e., in a rejection solution over a rectangular domain more than 30% of all of the points evaluated would be rejected.

Exercise 1.4

Statement of the problem

4. Using the random number generator 'fltrn' model Buffon's experiment and obtain an estimate of π.

Solution I

Buffon's experiment is described in the *Primer* on pages 2-3. As a first approach to solving this exercise we consider what is perhaps the most obvious method of solution. We assume the lines in Buffon's experiment are spaced one unit of distance apart such that, in the notation of eqn 1.4P, d = 1. As suggested by the discussion on p. 3 of the *Primer*, we use $\lambda = d/2$, and thus $\lambda = 0.5$. With these parameters set, we choose an x at random in the interval (0,1), an angle α in the interval $(0,\pi/2)$, and determine whether x is in region L of Figure 1.2P; i.e., whether $x < \frac{1}{2}\lambda \sin\alpha$ or $x > 1 - \frac{1}{2}\lambda \sin\alpha$. If either of these conditions is true we add a value of one to a running tally of k, the number of times the needle intersects a line. After determining k for n throws of the needle, we calculate p according to eqn 1.2P. From p we can obtain an estimate of π using eqn 1.4P. For the selection of the angle α we use the transformation shown in Exercise 1.2, that is given by eqn 1.1, with $a = \pi/2$.

A Fortran program to calculate p, and thus to estimate π, using this approach is given in Table 1.4. The standard deviation of p is estimated using eqn 1.9P. With a sample of 10^6 throws we obtain $p = 0.31842 \pm 0.00047$, and thus we estimate $\pi \approx 3.14051 \pm 0.00464$. From Table 1.4P we see that our earlier estimate of π, obtained using the rejection technique to calculate the area of a quadrant of a unit circle and using the same number (10^6) of sample points, was 3.142096 ± 0.001642. That is, in some sense the rejection technique of Example 1.2 appears to be a "better" method for estimating π than Buffon's method, assuming the run times of the two calculations are comparable.

Table 1.4. Fortran Program for Exercise 1.4, Solution I

```
      DOUBLE PRECISION fltrn, pi
      pi=3.14159265
      WRITE(*,*)
    2 WRITE (*,'(1X,A\)') ' Enter number of throws '
      READ (*,*) nmax
      IF(nmax.eq.0) STOP
      nscore = 0
      DO 1 i = 1,nmax
         x = fltrn()
         alpha = pi*fltrn()/2.
         y = SIN(alpha)/4.
         IF(x.lt.y.or.x.gt.1.-y)nscore = nscore+1
    1 CONTINUE
      prob = FLOAT(nscore)/FLOAT(nmax)
      sig = SQRT(prob*(1.-prob)/FLOAT(nmax))
      calcpi = 1./prob
      WRITE (*,'(1X,A\)') ' k/n and sig '
      WRITE(*,*)prob, sig
      WRITE(*,'(1x,a\)')' pi '
      WRITE(*,*)calcpi
      GOTO 2
      END
```

Solution II

The program of Table 1.4 correctly models Buffon's experiment and thus is a solution to the exercise. However, the formulation of that solution is somewhat unsatisfactory because it requires the user to supply a value of π in order to perform the calculation. This introduces circulatory logic into the calculation that might be deemed undesirable. To avoid such circulatory logic it is possible to begin with a "rough" estimate for π, and then iterate to obtain a better value. A program to do this is shown in Table 1.5. This program uses an initial guess of $\pi \approx 3.0$. With this initial guess, about five iterations of 10^6 throws each are required in order to determine $\pi \approx 3.14$. This iterative approach to the solution of Buffon's experiment eliminates the need for exact *a priori* knowledge of π and is thus, perhaps, more aesthetically pleasing than Solution I. One could, of course, examine the convergence of this iterative scheme to see whether it is stable and whether the correct value of π would be obtained independent of the accuracy of the initial guess.

Table 1.5. Fortran Program for Exercise 1.4, Solution II

```
      DOUBLE PRECISION fltrn, pi
      pi=3.0
      WRITE(*,*)
    2 WRITE (*,'(1X,A\)') ' Enter number of throws '
      READ (*,*) nmax
      IF(nmax.eq.0) STOP
      nscore = 0
      DO 1 i = 1,nmax
         x = fltrn()
         alpha = pi*fltrn()/2.
         y = SIN(alpha)/4.
         IF(x.lt.y.or.x.gt.1.-y)nscore = nscore+1
    1 CONTINUE
      prob = FLOAT(nscore)/FLOAT(nmax)
      sig = SQRT(prob*(1.-prob)/FLOAT(nmax))
      pi = 1./prob
      WRITE (*,'(1X,A\)') ' k/n and sig '
      WRITE(*,*)prob, sig
      WRITE(*,'(1x,a\)')' pi '
      WRITE(*,*)pi
      GOTO 2
      END
```

Solution III

There is another technique by which the Buffon problem can be solved that requires the user to make no assumptions at all about the value of π. In the Buffon problem the angle α is determined by the physical orientation of the needle, and this orientation can be sampled directly using the rejection technique of Example 1.2. In that example we pick a point (x,y) randomly within the positive quadrant of a unit circle centered at the origin by requiring $x^2+y^2 \leq 1$. Here we could pick any radius within which to select a point but it is convenient to use a unit circle. A line between the origin and such a random point makes a random angle with the X axis and therefore corresponds to the random angle α in Buffon's experiment, as shown in Figure 1.1P. The sine of this angle is $y/\sqrt{(x^2+y^2)}$.

A program for applying this rejection technique to the Buffon problem is shown in Table 1.6. With 10^6 samples this program gives p $\approx$ 0.31828 $\pm$ 0.00047 and thus we estimate $\pi \approx 3.1419 \pm 0.0046$. As expected, these results are similar to those obtained with the program of Table 1.4.

Table 1.6. Program for Exercise 1.4, Solution III

```
      DOUBLE PRECISION fltrn
      WRITE(*,*)
    2 WRITE (*,'(1X,A\)') ' Enter number of throws '
      READ (*,*) nmax
      IF(nmax.eq.0) STOP
      nscore = 0
      DO 1 i = 1,nmax
    5   x=fltrn()
        y=fltrn()
        rsq=x**2+y**2
        IF(rsq.GT.1.0)GOTO 5
        SinAlpha=y/SQRT(rsq)
        x = fltrn()
        y = SinAlpha/4.
        IF(x.lt.y.or.x.gt.1.-y)nscore = nscore+1
    1 CONTINUE
      prob = FLOAT(nscore)/FLOAT(nmax)
      sig = SQRT(prob*(1.-prob)/FLOAT(nmax))
      pi = 1./prob
      WRITE (*,'(1X,A\)') ' k/n and sig '
      WRITE(*,*)prob, sig
      WRITE(*,'(1x,a\)')' pi '
      WRITE(*,*)pi
      GOTO 2
      END
```

Discussion

It is apparent from the solutions offered for the exercises in this chapter that the rejection technique is a powerful Monte Carlo sampling tool. While other sampling formulations for which closed-form solutions exist, such as those that are introduced in Chapter 2 of the *Primer*, may offer greater calculational efficiency than rejection, the effort required to implement such formulations may not justify their use. In addition, rejection is the preferred method in many sampling applications. We encounter such a situation when considering the thermal scattering of neutrons and the need for sampling from eqn 9.51P.

Chapter 2

Monte Carlo Sampling Techniques

Exercise 2.1

Statement of the problem

1. Using eqns 2.8P and 2.9P derive a sampling scheme for selecting the radii of points uniformly distributed over
   a. a disk of radius r_o
   b. an annulus of inner radius r_i and outer radius r_o
   c. a spherical shell of inner radius r_i and outer radius r_o.

Monte Carlo calculations often require the user to select source particles from a particular spatial distribution. The current exercises illustrate three common types of uniform spatial distributions.

Solution Part a

Assume the disk lies in the X-Y plane, and is centered on the Z axis. The probability of a source particle being born at a radius r_s can be obtained from the cumulative distribution function

$$P(r_s) = \frac{\int_0^{r_s} 2\pi r dr}{\int_0^{r_o} 2\pi r dr} = \frac{r_s^2}{r_o^2} \tag{2.1}$$

Since $P(r_s) \leq 1$ for $0 \leq r_s \leq r_o$, we can define the radius of a random point r_s by setting $P(r_s) = \xi$, where ξ is a random number over the interval (0,1). Thus the radius of a random start point can be found by obtaining a random number, ξ, and calculating

$$r_s = \sqrt{\xi r_o^2}$$

Solution Part b

Assume the annulus is defined by two right circular cylinders centered on the Z axis. The probability of a particle being born at radius r_s, $r_i \leq r_s \leq r_o$, can be obtained from the cumulative distribution function

$$P(r_s) = \frac{\int_{r_i}^{r_s} 2\pi r dr}{\int_{r_i}^{r_o} 2\pi r dr} = \frac{r_s^2 - r_i^2}{r_o^2 - r_i^2}$$

Thus as in part a of this exercise, the radius of a random start point can be found by obtaining a random number, ξ, and setting $P(r_s) = \xi$, for which we determine

$$r_s = \sqrt{\xi\left(r_o^2 - r_i^2\right) + r_i^2}$$

Solution Part c

Assume the spherical shell is centered at the origin. Then the probability that a source particle will be born at a radius r_s, $r_i \leq r_s \leq r_o$, can be obtained from the cumulative distribution function

$$P(r_s) = \frac{\int_{r_i}^{r_s} 4\pi r^2 dr}{\int_{r_i}^{r_o} 4\pi r^2 dr} = \frac{r_s^3 - r_i^3}{r_o^3 - r_i^3}$$

As before the radius of a random start point can thus be found by obtaining a random number, ξ, and calculating

$$r_s = \left\{r_i^3 + \xi\left(r_o^3 - r_i^3\right)\right\}^{1/3}$$

Discussion

Finding the radius of a random point as measured from an axis of symmetry in some geometric distribution, which is required by the preceding exercise, is but the first step in specifying the Cartesian coordinates of the random point. Such coordinates must account for the location of the point in three-dimensional space and must therefore be determined based on the actual location and orientation of the geometric distribution. This frequently requires selection of a random unit vector and may require coordinate translation and rotation. The former is discussed in Example 3.1. Rotational transformations are discussed in Section 4.2P.

There is an alternative method for selecting points randomly on a disk or randomly within a spherical volume. This well-known technique is applicable to any random variable that has a distribution of the form x^n, where n is an integer. The method is to select n random numbers and accept the largest for evaluating the random variable. That is, for the disk problem in part a of Exercise 2.1 the distribution we wish to sample is given by eqn 2.1; i.e., $P(r_s)$ is distributed as r_s^2. We can thus select a radius from the points distributed uniformly over a disk of radius r_o by selecting the larger of two random numbers and multiplying that value by r_o. Likewise we can select from the distribution of the radii of points uniformly distributed over a spherical volume, as given by eqn 2.17P, by selecting the largest of three random numbers and multiplying that value by the radius of the sphere.

The basis for this technique can easily be shown. Given the selection of two random numbers ξ_1 and ξ_2 in (0,1), the probability that at least one of the numbers (alternately, the largest of the two numbers, ξ_L) will be greater than $x \in (0,1)$ is given by

$$\begin{aligned}\Pr(\xi_L > x) &= \Pr(\xi_1 > x) + \Pr(\xi_1 \le x) \times \Pr(\xi_2 > x) \\ &= (1 - x) + x(1 - x) \\ &= 1 - x^2\end{aligned}$$

Now consider the nth case. Assuming that, given the selection of n random numbers in (0,1), the probability that one of the numbers is greater than $x \in (0,1)$ is given by $1\text{-}x^n$ then the probability that n+1 random numbers will have a number greater than x is given by

$$\begin{aligned} Pr(\xi_L > x) &= Pr(\xi_{1\to n} > x) + Pr(\xi_{1\to n} \le x) \times Pr(\xi_{n+1} > x) \\ &= (1 - x^n) + x^n(1 - x) \\ &= 1 - x^{n+1} \end{aligned}$$

where $\xi_{1\to n}$ refers to the n random numbers smaller than ξ_L. By induction, for all n > 0, given the selection of n random numbers in (0,1), the probability that the largest of the numbers will be greater than $x \in (0,1)$ is given by

$$Pr(\xi_L > x) = 1 - x^n$$

Alternately, the probability that the largest of the random numbers will be less than or equal to x is given by

$$Pr(\xi_L \le x) = x^n$$

Therefore, given the selection of n random numbers, the distribution of the largest random number is given by x^n.

Exercise 2.2

Statement of the problem

2. Using the rejection technique derive a sampling scheme to select a random value for $\cos\theta$, assuming that θ is uniformly distributed from $[0,2\pi]$. The sampling scheme should not use a function to calculate the cosine of an angle. Calculate an average value for the cosine over $[0,\pi/2]$ and for $[0,2\pi]$. For each case, estimate the variance of the distribution, the variance of the estimate of the mean, and the fractional standard deviation (fsd). Show why the fsd is not a useful measure for the average of the cosine over $[0,2\pi]$. [Hint: If $R = x^2 + y^2 \le 1$, $\cos\theta = x/\sqrt{R}$.]

Solution

The procedure implied by the hint for sampling from the cosine of an angle without evaluating a function is to sample points within a unit circle. As in Exercise 1.4, solution c, we could pick any radius within which to select a point but it is convenient to use a unit circle. Again for convenience we assume this circle lies in the X-Y plane and is centered at the origin. We want to find a random point within this unit circle using the rejection technique, as is done in Example 1.2, calculate the square of the radius R of a circle centered at the origin that passes through the selected point, and obtain the cosine from $x/\sqrt{R}$. Since the points found by this procedure are uniformly distributed within the unit circle, the resulting values for θ are also uniformly distributed from $[0,2\pi]$.

A program to select 10^6 random cosines, and the squares of these cosines, using this approach is given in Table 2.1. Having obtained the cosines, the coding in Table 2.1 then proceeds to determine the average cosine, the standard deviation of the samples taken, and the fractional standard deviation of the mean. The program is similar to that of Table 1.3P; however, because we want to cover the complete circle $[0,2\pi]$, we must allow x and y to take on negative values. That is, our cell is bounded by $-1 < x < 1$ and $-1 < y < 1$ rather than $0 < x < 1$ and $0 < y < 1$ as is the case in Example 1.2. Rejecting points outside the unit circle (or circle of any radius) ensures the result $x/\sqrt{R}$ will be uniformly distributed over the circle.

The average cosine over $[0,2\pi]$ is obtained by summing the 'costh' results from the program of Table 2.1 and dividing by the number of values summed. Obtaining the average cosine over $[0,\pi/2]$ requires that the coding be changed to sample from only the positive quadrant of the unit circle; i.e., the sampling in this case is identical to that of Table 1.3P. The Fortran required for sampling over $[0,\pi/2]$ is indicated in Table 2.1 by the incorporated comments. Finally, the estimate of the variance of the result for these two cases is found by computing and summing the squares of the variable 'costh' and using eqns 2.43P and 2.44P. The fractional standard deviation is given by the square root of the variance divided by the calculated mean cosine.

Although it is not required for the solution of the exercise, the procedure that is described beginning with eqn 2.30P on p. 33 of the *Primer* can be followed to solve eqn 2.29P and obtain an analytical solution for the variance of the distribution over the intervals $[0,2\pi]$ and $[0,\pi/2]$. To obtain the variances we use

$$\int_{\alpha}^{\beta} \cos\theta d\theta = \sin\beta - \sin\alpha$$

and

$$\int_{\alpha}^{\beta} \cos^2\theta d\theta = \frac{\beta-\alpha}{2} + \frac{\sin 2\beta - \sin 2\alpha}{4}$$

Table 2.1. Fortran Program for Exercise 2.2

```
      REAL(8) fltrn
      nmax = 1000000
      costot = 0.
      Costot2 = 0.
      DO 1 I = 1, nmax
!    following two lines calculate avg cos over [0,2*pi]
  2   x = 1.-2.*fltrn()
      y = 1.-2.*fltrn()
!    to calc avg cos over [0,pi/2] replace prev two lines with
! 2     x = fltrn()
!       y = fltrn()
      rsq = x**2 + y**2
      IF(rsq.gt.1.)GOTO 2
      costh = x/SQRT(rsq)
      costot = costot + costh
      costot2 = costot2 + (costh)**2
 1    CONTINUE
      cosav = costot/FLOAT(nmax)
      cosav2 = costot2/FLOAT(nmax)
      sig = SQRT((cosav2-cosav**2)/FLOAT(nmax))
      fsd = sig/cosav
      WRITE(*,*)cosav,sig,fsd
      STOP
      END
```

The average value for the cosine over an interval [a,b] is given by

$$V = \frac{1}{(b-a)}\int_a^b \cos\theta d\theta$$

In the first case, for $\theta \in [0,2\pi]$, we use

$$f_1(x) = \begin{cases} \dfrac{1}{2\pi}, & 0 \le x \le 2\pi \\ 0, & \text{otherwise} \end{cases}$$

and

$$V_1(x) = \cos x$$

where the notation follows that of Chapter 2 of the *Primer*. Thus from eqn 2.23P we find

$$E(V_1) = \int_{-\infty}^{+\infty} V_1(x) f_1(x) dx = \frac{1}{2\pi}\int_0^{2\pi} \cos\theta d\theta = \frac{\sin 2\pi - \sin 0}{2\pi} = 0 \qquad (2.2)$$

and

$$E(V_1^2) = \frac{1}{2\pi}\int_0^{2\pi} \cos^2\theta d\theta = \frac{1}{2}$$

For the second case, $\theta \in [0,\pi/2]$, we use

$$f_2(x) = \begin{cases} \frac{2}{\pi}, & x \in \left[0, \frac{\pi}{2}\right] \\ 0, & \text{otherwise} \end{cases}$$

and

$$V_2(x) = \cos x$$

Thus

$$E(V_2) = \int_0^{\pi/2} \frac{2}{\pi}\cos\theta d\theta = \frac{2}{\pi} \approx 0.63662 \qquad (2.3)$$

and

$$E(V_2^2) = \int_0^{\pi/2} \frac{2}{\pi}\cos^2\theta d\theta = \frac{1}{2}$$

In the first case the variance of the distribution is given by

$$\text{var}(V_1) = E(V_1^2) - [E(V_1)]^2 = 0.5$$

while for the second case we have

$$\text{var}(V_2) = \frac{1}{2} - \frac{4}{\pi^2} \approx 0.09472$$

Running the Fortran program of Table 2.1 with nmax = 10^6 and the default start random number in 'fltrn' for the two intervals gives for $\theta \in [0,2\pi]$

$$\langle\cos\theta\rangle_1 = -5.468\times10^{-4}, \ \sigma_1 = 7.074\times10^{-4}$$

and for $\theta \in [0,\pi/2]$

$$\langle\cos\theta\rangle_2 = 0.636671, \quad \sigma_2 = 3.0718\times10^{-4}$$

By eqn 2.2 the correct result for the first case is zero, and by eqn 2.3 the correct result for the second case is about 0.63662. The standard deviation for the first case with 10^6 samples should be about $\sqrt{[\mathrm{var}(V_1)]}/1000 \approx 7.0711\times10^{-4}$, while that for the second case should be about 3.0776×10^{-4}. The calculation gives an fsd for the first case of −1.29 and an fsd for the second case of 0.00048. Obviously, because the true mean is zero, the fsd does not provide a good representation of the uncertainty in the Monte Carlo result for the first case.

Exercise 2.3

Statement of the problem

3. Repeat Exercise 3 of Chapter 1 using stratified sampling. Segment the integral into several different numbers of strata of equal size and, for the same total number of start particles, show how the standard deviation changes with the number of strata.

Solution

Exercise 1.3 requires us to determine the value of the integral

$$I = \int_{0.25}^{4} \Phi(x)dx$$

where Φ is the normal distribution. In Exercise 1.3 we obtain the result $I \approx 0.401207 \pm 0.000047$ using unbiased Monte Carlo with 10^8 points. By modifying the program used in Example 2.4, Table 2.7P, we can implement stratification in our Monte Carlo calculation of the above integral. Only the function to be integrated (line 5) and the limits (line 7) need to be changed from the program in Table 2.7P, as shown in Table 2.2. The subroutine shown in Table 2.8P is also used in this exercise.

Running the program of Table 2.2 with 10^6 points and different numbers of strata gives the results shown in Table 2.3. We see that the answer is unchanged to four significant digits for five strata or more, but the standard deviation continues to decrease almost linearly with the number of strata.

Table 2.2. Fortran Program for Exercise 2.3

```
!  Program to implement stratification
!  arrays are dimensioned to handle up to 100 strata
     DOUBLE PRECISION dx,a,b,score,delta,f,fltrn,delta2
     COMMON/scores/score,n
     f(dx)=DEXP(-dx*dx/2.0d0)/DSQRT(2.0d0*3.14159265d0)      ! function to be integrated
     OPEN(unit=8,file='output.txt')
     a = 0.25d0; b = 4.0d0                            ! a,b are lower,upper limits of integral
     nsamples = 1000000                               ! nsamples-number of samples to be performed
     WRITE(8,13)nsamples,a,b
13   FORMAT(' no of samples=',I8,'   Lower Limit=',f10.6,'   Upper Limit=',f10.6)
     WRITE(*,15)
15   FORMAT(' Please enter the number of strata up to 100.')
     READ(*,*)nstrata                                 ! number of strata to be used
     IF((nstrata.LT.1).OR.(nstrata.GT.100))nstrata=100
     CALL stats
     delta=(b-a)/nstrata                             ! delta x value for each stratum
     delta2=(b-a)                                    ! delta x for unstratified calculation
     DO 100 j=1,nsamples
       dx=fltrn()                                    ! fltrn() generates a random number in (0,1)
       n=1+nstrata*dx                                ! calculate the strata associated with dx
       dx=a+(b-a)*dx; score=delta*f(dx)              ! calculate score at dx
       CALL statlp                                   ! store statistics for score
       score=delta2*f(dx)                            ! calculate score for unstratified case
       n=nstrata+1                                   ! store unstratified scores in last strata
       CALL statlp                                   ! store statistics for unstratified case
100  CONTINUE
     n=nstrata
     CALL statend                                    ! calculate and print results
     STOP
     END
```

Table 2.3. Results for Exercise 2.3

Number of Strata	Value of Integral	Standard Deviation
1	0.401021	0.471-3
2	0.401356	0.294-3
5	0.401235	0.110-3
7	0.401178	0.789-4
10	0.401229	0.554-4
20	0.401220	0.277-4
40	0.401247	0.138-4
60	0.401257	0.923-5
80	0.401276	0.692-5
100	0.401250	0.553-5

From these results we can estimate that the use of about twelve strata sampled with 10^6 points should give a result with roughly the same accuracy as that obtained in Exercise 1.3 using 10^8 points without stratification. When we consider that the stratified result is achieved in a run time that is approximately a factor of 100 less than the non-stratified result (i.e., by using 10^6 points instead of 10^8 points) it is clear that we have obtained a significant gain in efficiency as defined in the discussion of Exercise 1.2.

Discussion

It can be seen from Table 2.3 that when the number of strata is increased from 10 to 100 the standard deviation appears to be decreased by a factor of ten. By modifying the program to allow more than 100 strata we can examine this result further. The results of running the calculation with 1000 and 10000 strata show that the standard deviation continues to decrease by a factor of ten each time the sample size is increased by a factor of ten. This result is expected for smooth functions where the strata are sufficiently small such that the variation of the function across each individual stratum is essentially linear. The effects of stratification on linear functions can easily be demonstrated by modifying the program in Table 2.2 to evaluate the area under a straight line, for example by setting 'f(dx) = dx.'

Exercise 2.4

Statement of the problem

4. Calculate the value of the integral

$$\int_0^{2\pi} \frac{\sin\theta d\theta}{\theta}$$

using analog Monte Carlo. Implement several different numbers of strata and determine the gains in efficiency produced thereby. Define a modified pdf that appears to offer an improvement in efficiency and recalculate the integral. How much improvement was obtained?

Solution

Analog and stratified Monte Carlo solutions for the desired integral can be obtained in a manner similar to that used in the solution of Exercise 2.3. That is, the programs given in Tables 2.7P and 2.8P are used, with the function definition in line 5, and the limits in line 7, of Table 2.7P changed to reflect the current integral. Running such a code modified from Example 2.4, for the default number of 10^6 samples and for various numbers of strata, gives the results shown in Table 2.4 (where the case of one stratum is the analog result). As in Exercise 2.3 we see that the standard deviation decreases almost linearly with an increase in the number of strata.

Table 2.4. Analog Results for Exercise 2.4

Number of Strata	Value of Integral	Standard deviation
1	1.416401	2.71-03
2	1.418481	1.46-03
10	1.418313	3.02-04
20	1.417967	1.51-04
40	1.418091	7.58-05
60	1.418116	5.06-05
80	1.418249	3.79-05
100	1.418087	3.03-05

There are many ways to define a modified probability density function (pdf) as a means to increase the efficiency of the present calculation. We consider one simple modification. Figure 2.1 shows a plot of the integrand $f(\theta) = \sin\theta/\theta$ as a function of the independent variable θ. Let us define a straight line $y(\theta)$ from the point (0,1) to the point $(2\pi,0)$; i.e.,

$$y = 1 - \frac{\theta}{2\pi}, \quad 0 < \theta < 2\pi$$

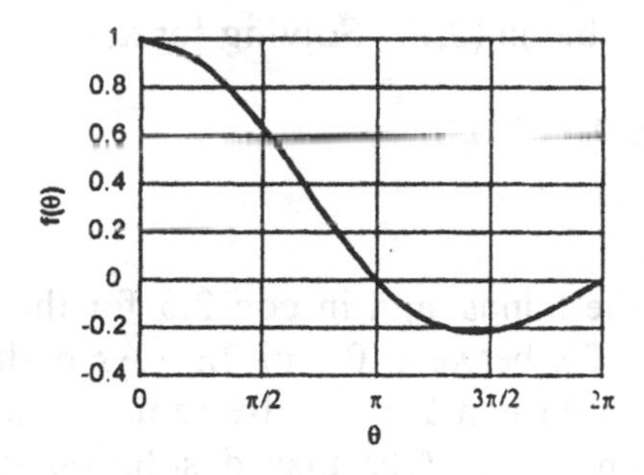

Figure 2.1. Function of Exercise 2.4

Obviously this straight line lies closer to the value of the function, on average, than a line parallel to the θ-axis. Therefore we would expect that using this line as the basis for a modified pdf would improve the efficiency of our calculation.

Using this line, with appropriate normalization, we can define a modified pdf g(x). That is, following the development in Section 2.5.2P,

$$V(x) = \frac{2\pi \sin x}{x}$$

$$f(x) = \frac{1}{2\pi} \ , \ x \in (0, 2\pi)$$

$$g(x) = \frac{1 - \dfrac{x}{2\pi}}{\pi} \ , \ x \in (0, 2\pi) \tag{2.4}$$

$$V'(x) = \frac{V(x)f(x)}{g(x)} = \frac{\pi \sin x}{x\left(1 - \dfrac{x}{2\pi}\right)}$$

We want to determine the expected value of V′, choosing x from the modified distribution g. Thus we calculate

$$\int_0^x g(t)dt = \frac{x - \dfrac{x^2}{4\pi}}{\pi} = \xi$$

where ξ is a random number in (0,1). Solving for x,

$$x = 2\pi\left[1 \pm (1-\xi)^{1/2}\right] \tag{2.5}$$

We need to select the minus sign in eqn 2.5 for the present solution in order to obtain values of x between 0 and 2π. As is discussed in Section 2.2.1P, the factor $1 - \xi$ in eqn 2.5 can be replaced by ξ since both are randomly distributed on (0,1). Our biased solution is then obtained by modifying the Fortran of Example 2.5, shown in Table 2.9P, to reflect the expression for g(x) given by eqn 2.4. The resulting program is shown in Table 2.5. The notation in this program uses f(x) and g(x) to represent the unbiased function V(x) and the biased function V′(x), respectively.

The result obtained from running this program with a sample size of 10^6 and the modified pdf is 1.41921 ± 0.00206. This compares with the unbiased result of 1.41640 ± 0.00271 and the stratified result of Table 2.4, with 100 strata, of 1.418087 ± 0.000030. The present selection for a biased distribution, which is neither unique nor optimum, appears to have improved the accuracy of the result compared with the unbiased calculation. It is easy to show that it would take about 1.7 times more samples in the unbiased case to duplicate the standard deviation obtained in the biased calculation.

Table 2.5. Fortran for Biased Solution to Exercise 2.4

```
      DOUBLE PRECISION x,a,b,score,sumf,sumfsq,stdev,f,r,sumg,sumgsq,g,fltrn,pi,tpi
      PARAMETER (pi=3.14159265,tpi=2.0d0*pi)
      f(x)=tpi*DSIN(x)/x                    ! f(x) is the unbiased function, V(x)
      g(x)=pi*DSIN(x)/(x*(1.0d0-x/tpi))     ! g(x) is the biased function, V'(x)
      OPEN(unit=6,file='output.txt')
      a=0.d0; b=tpi                         ! a,b are lower and upper limits of integral
      sumf=0.0; sumfsq=0.0; sumg=0.0d0; sumgsq=0.0d0          ! Initialize variables
      nsamples=1000000                       ! nsamples is the number of samples
      WRITE(6,13)nsamples,a,b
13    FORMAT(' no of samples=',i8,'   Lower Limit=',f10.6,'   Upper Limit=',f10.6)
      DO 100 j=1,nsamples
        r=fltrn()                            ! fltrn() generates a random number in (0,1)
        x=a+(b-a)*r                                            ! pick unbiased x
        score=f(x); sumf=sumf+score; sumfsq=sumfsq+score**2    ! score unbiased function
        x=tpi*(1-DSQRT(r))                                     ! pick biased x
        score=g(x); sumg=sumg+score; sumgsq=sumgsq+score**2    ! score biased function
100   CONTINUE
      j=nsamples
      stdev = DSQRT(DABS(sumfsq/j - (sumf/j)**2))
      WRITE(6,14)j,sumf/j,stdev,stdev/SQRT(FLOAT(j))           ! unbiased results
      stdev = DSQRT(DABS(sumgsq/j - (sumg/j)**2))
      WRITE(6,14)j,sumg/j,stdev,stdev/SQRT(FLOAT(j))           ! biased results
 14   FORMAT (1x,i7,f14.8,f14.8,f14.8)
      STOP
      END
```

Discussion

The modified distribution function selected for this exercise, given by eqn 2.4, reduces the standard deviation of the result by about 25% for the same number of samples, compared with the unbiased calculation. While not huge, this reduction is nonetheless significant.

We have not as yet defined a quantitative expression for the efficiency of a Monte Carlo calculation. Reducing the variance in a result does not imply that a gain was made in efficiency since it may have taken longer to obtain the improved result than to obtain the original result. Knowing that the standard deviation decreases inversely with the square root of the number of samples, one can always improve a Monte Carlo result by including more samples. However, the run time of a calculation should be taken into account in determining the efficiency of a calculation. A practical formula for efficiency is introduced in Chapter 6 of the *Primer*.

Chapter 3

Monte Carlo Modeling of Neutron Transport

Exercise 3.1

Statement of the problem

1. By converting the collision density distribution of eqn 3.4P into a pdf, show that the "expected value" of the distance traveled is $1/\Sigma_t$.

Solution

Eqn 3.4P is the exponential distribution,

$$n(x) = n(0)e^{-\Sigma_t x}$$

The pdf for this distribution is obtained by finding the appropriate normalization. Consider

$$f(x) = \Sigma_t e^{-\Sigma_t x}, 0 \le x < \infty \tag{3.1}$$

Then f(x) is a pdf since $f(x) \ge 0$ and $\int f(x)dx = 1$ over the limits $0 \le x < \infty$. A random point can be selected from the distribution f(x) by setting the corresponding cumulative distribution function equal to a random number ξ over the range (0,1). That is,

$$F(x) = \int_0^x f(x)dx = \xi$$

Substituting eqn 3.1 for f(x) gives

$$\xi = \int_0^x \Sigma_t e^{-\Sigma_t y} dy = 1 - e^{-\Sigma_t x}$$

Solving for x, and taking into account the fact that both ξ and $1 - \xi$ are random numbers evenly distributed over (0,1), gives

$$x = \frac{-\ln \xi}{\Sigma_t} \qquad (3.2)$$

which is just eqn 2.14P.

The Fortran in Table 3.1 uses eqn 3.2 to sample for x and determines the mean free path (mfp) for neutron transport in a material with a total cross section $\Sigma_t = 1$. The standard deviation is also estimated using eqn 2.43P. Using 10^4 samples, this code gives the result 0.9993 ± 0.0103 for the mean flight path between collisions. Because we set $\Sigma_t = 1$ the correct result is 1.0. Obviously by our definition of f(x) above, and by eqn 2.14P, if we choose Σ_t to be some other value than 1.0 we would multiply the value of x in the Fortran by $1/\Sigma_t$ and hence our result would be $0.9993/\Sigma_t \approx 1/\Sigma_t$, which is the expected value for the particle flight path.

Table 3.1. Fortran Program for Exercise 3.1

```
      DOUBLE PRECISION fltrn
      nmax = 10000
      sum = 0.
      sumsq = 0.
      sigmat = 1.0
      DO 1 i = 1,nmax
        x = -DLOG(fltrn())/sigmat
        sum = sum + x
        sumsq = sumsq + x**2
1     CONTINUE
      sum = sum/FLOAT(nmax)
      sumsq = sumsq/FLOAT(nmax)
      sig = SQRT((sumsq-sum**2)/FLOAT(nmax))
      WRITE(*,*)sum,sig
      STOP
      END
```

Discussion

An alternative approach to this problem involves deriving the pdf for the exponential distribution from first principles. The probability dP that a neutron will travel a distance x without suffering a collision in a uniform, homogeneous material with a macroscopic total cross section Σ_t, and then will experience a collision in dx about x is

$$dP = \Sigma_t e^{-\Sigma_t x} dx$$

The probability dP is the product of Σ_t, which is the probability per unit path length of a neutron experiencing a collision in the material in question; times the path length dx; times $e^{-\Sigma_t x}$, which is the probability that the neutron will reach the point x without suffering a collision. Let

$$f(x) = \frac{dP}{dx} = \Sigma_t e^{-\Sigma_t x}$$

Then, as before, we select a distance by setting the cumulative distribution function equal to a random number over the interval (0,1),

$$F(x) = \int_0^x f(x)dx = \xi$$

Exercise 3.2

Statement of the problem

2. Start the source particles in Example 3.2 at random locations and directions (i.e., ignore the symmetry of the example geometry) and show that the results do not change, within the statistical accuracies of the calculations.

Solution

In Example 3.2 we calculate the uncollided photon leakage from a homogeneous spherical source of gamma rays. To simplify the determination of the start locations of the source particles in that example, we use the symmetry of the geometry to argue that we could start all particles on the +Z axis. In the current exercise we select start points uniformly throughout the spherical source region. This requires selecting a random vector isotropic over a 4π solid angle, and a radius r that is distributed according to the cumulative distribution function described in Example 2.1, eqn 2.16P, for uniform sampling inside a sphere. The intersection of the vector and the sphere of radius r, when both are centered at the origin, defines the desired random source point.

We can use subroutine 'Isoout,' shown in Table 3.2P, to define a random vector centered at the origin of a Cartesian coordinate system. This subroutine calculates the direction cosines of a random unit vector Ω. Moving from the origin of our coordinate system a distance r in the direction Ω defines a random point on the spherical surface of this radius. By eqns 3.29P the coordinates of this random point are given by

$$x = r\sin\theta\cos\varphi$$

$$y = r\sin\theta\sin\varphi$$

$$z = r\cos\theta$$

As in Example 3.2, the radius of the start point, r, can be found by using eqn 2.18P. Suitably modifying the program of Table 3.4P gives the program shown in Table 3.2. This program calls the version of 'Isoout' shown in Table 3.2P.

Table 3.2. Fortran Program for Exercise 3.2

```
      DOUBLE PRECISION fltrn
      WRITE(*,*)
New_Case_Loop: DO      !  The loop terminates if NMAX=0
      WRITE(*,'(1X,A\)')' Enter number of start particles '   ! Read input
      READ(*,*)NMAX                                           ! Number of particles
      IF (NMAX.LE.0) EXIT New_Case_Loop
      WRITE(*,'(1x,a\)') ' Enter sphere radius in mfp '
      READ(*,*) r;          rsq = r*r                         ! Radius of sphere
      WRITE(*,'(1X,A\)')' Enter start random number seed '
      READ(*,*)ISEED                                          ! Random number seed
      CALL rndin(ISEED)                                       ! Set start random number
      nr = 0
      DO 300 n=1,NMAX   !     Begin particle tracking -- loop over NMAX particles
        rad = r*fltrn()**0.333333333          ! pick uniformly from sphere
        CALL Isoout(us,vs,ws)
        x=rad*us;  y=rad*vs;  z=rad*ws
        CALL Isoout(u,v,w)
        d = -DLOG(fltrn())                    ! Get a flight path distance
        xc = x+d*u                            ! distance to collision along x-axis
        yc = y+d*v                            ! distance to collision along y-axis
        zc = z+d*w                            ! distance to collision along z-axis
        rcsq = xc*xc+yc*yc+zc*zc              ! square of radius to collision point
        IF(rcsq.GT.rsq) nr=nr+1               ! did the particle escape sphere?
  300 CONTINUE
      prob = FLOAT(nr)/FLOAT(nmax)      !  Calculate expected escape probability
      StDev=SQRT(prob-prob**2)/SQRT(FLOAT(nmax)) ! Standard deviation from eqn 2.29 for
      WRITE (0,13)nr, prob, StDev       !    estimate of mean for a binomial distribution
   13 FORMAT(' Particles escaping =',I9,', probability of escape =',f11.4,&
       /,'standard deviation = ',f12.5)
END DO New_Case_Loop
      STOP;     END
```

Running this code with 10^6 start particles for r = 1, 2, and 5 we obtain leakage probabilities of 0.5261 ± 0.0005, 0.3318 ± 0.0005, and 0.1470 ± 0.0004. Comparing these results with those of Table 3.5P we see that the two methods for positioning start particles give the same answers, within the statistical uncertainties of the results.

Exercise 3.3

Statement of the problem

3. Consider the slab with the bubble in Example 3.5.
   a. Reduce the size of the bubble and show that the result approaches that of Example 3.3 as the bubble radius goes to zero.
   b. Modify the program such that the incident pencil beam is tangent to the bubble. How do the results compare with those of Examples 3.3 and 3.5?

Solution Part a

To solve part a of this exercise we start with the Fortran of Example 3.5. Subroutines 'Isoout' and 'Output,' shown in Tables 3.2P and 3.8P, are included without modification. In the routines of Tables 3.11P and 3.12P, which are also not changed, the radius of the bubble is given by the variable 'R0.' To vary the radius we add this variable as an input parameter to the version of subroutine 'Input,' shown in Table 3.7P. The modified subroutine 'Input' is shown in Table 3.3.

Table 3.3. Modified Subroutine 'Input' for Exercise 3.3, Part a

```
SUBROUTINE INPUT                           ! Read input data
COMMON/scoring/NA,NR,NL,NS
COMMON/geometry/R0,Z1,Z2                          ! new statement for Exer 3.3a
WRITE(*,'(1X,A\)')' Enter number of start particles '
READ(*,*)NS
IF (NS.LE.0) STOP
WRITE(*,'(1x,a\)')' enter seed '
READ(*,*)iseed
CALL rndin (iseed)
WRITE(*,'(1x,a\)')' Enter bubble radius '  ! new statement for Exer 3.3a
READ(*,*)r0                                ! new statement for Exer 3.3a
NA = 0;  NR = 0;  NL = 0               ! Initialize event counts
RETURN
END
```

The results obtained by running this code for 10^5 start particles and various values of 'R0' are shown in Table 3.4. The table shows the number of particles transmitted, reflected, and absorbed as the radius of the bubble approaches zero. The results for the case of a zero-radius bubble are identical to those that would be obtained when solving Example 3.3 for the case of 10^5 start particles with a random number seed of one.

Table 3.4. Results for Exercise 3.3, Part a

bubble radius	transmitted	reflected	absorbed
0.5	20903±129	48700±158	30397±145
0.25	15610±115	50539±158	33851±149
0.1	13711±109	51250±158	35039±151
0.	12766±106	51539±158	35695±151

Solution Part b

The problem does not state whether the modified beam must be normally incident on the slab. Therefore this part of the exercise can be interpreted at least two ways. One may assume the positions of the beam and bubble remain the same but the angle of incidence of the beam changes such that the extension of the beam into the slab is tangent to the bubble. Alternatively one may assume the beam remains normally incident on the slab and either the position of the beam or the position of the bubble changes such that the extension of the beam is tangent to the bubble. We consider only the second option.

In order to modify the relative positions of the incident beam and the bubble so that the beam is tangent to the bubble, either the position of the start particles or the location of the bubble (or both) could be changed. For the present calculation the position at which the beam is incident on the slab is changed from an x value of zero to an x value equal to the radius of the bubble. That is, the beam is still directed parallel to the +Z axis but it strikes the slab at the point (0.5,0) instead of the point (0,0) in the X–Y plane. This repositioning of the point of incidence of the beam is done by changing the second line after the 'DO 300' statement in Table 3.11P to set x = 0.5. The source point remains at the z value of negative 'Z1' because the dimensions of the slab in the Z direction have not changed.

Results obtained from running this modified code with 10^5 start particles are shown in Table 3.5 for source positions from x = 0 to x = 1.0, including the tangent position of x = 0.5. The results for the case x = 0 duplicate those of Example 3.5. The x = 1 results approach those of Example 3.3 in which there was no bubble. The results for the number of reflected particles varies between 48700 and 51336 per 10^5 start particles as the position of the beam varies between x = 0 and x = 1. That is, although the bubble is buried 1.5 mfp deep in the slab we see the effect of its location relative to the position of the incident beam on the number of particles that are reflected from the slab. The estimated uncertainty in the number of reflected particles, obtained using the expression for the standard deviation of a binomial distribution as coded in subroutine 'Output' of Table 3.8P (which is to be derived in the next exercise) is the same for all of the cases shown in Table 3.5.

Table 3.5. Results for Exercise 3.3, Part b

x	transmitted	reflected	absorbed
0.0	20903±129	48700±158	30397±145
0.1	20681±128	48735±158	30584±146
0.25	19271±125	49207±158	31522±147
0.5	13539±108	51134±158	35327±151
0.75	13285±107	51253±158	35462±151
1.0	13152±107	51336±158	35512±151

Exercise 3.4

Statement of the problem

4. The variances in the results of Example 3.3 were obtained assuming a binomial distribution in the estimates. This is done in subroutine 'Output' of Table 3.8P. Using eqn 2.29P derive the expression for the variance of a binomial distribution as encoded in Table 3.8P. Apply this variance estimate to the results of Example 1.2.

Solution

A binomial distribution is characterized by the fact that the score associated with an event is binary; i.e., the outcome of the event is always either yes or no, to which outcomes we can assign a score of one or zero. Thus the expected value of the scores, E(V), is equal to the probability of a score being one. If n_t events are scored, of which n_1 have a score of one and $n_t - n_1$ have a score of zero, the probability of scoring a one is

$$p = \frac{n_1}{n_t}$$

Because every non-zero score is equal to one, the square of every non-zero score is also equal to one. Hence,

$$E(V) = E(V^2) = p$$

and from eqn 2.29P,

$$\text{var}(V) = E(V^2) - [E(V)]^2 = p - p^2 = p(1-p) \tag{3.3}$$

The estimate of the variance of the mean of a Monte Carlo sample for a binomial distribution is then, from eqn 2.44P,

$$\text{var}(\langle V \rangle) = \frac{1}{n_t}\text{var}(V) = \frac{p(1-p)}{n_t} \tag{3.4}$$

The coding for estimating the standard deviations used in Table 3.4P makes use of the square root of this expression. In Table 3.8P, however, the output is the actual number of particles undergoing, for example, absorption, along with the associated standard deviation. If n_a is the estimate of the number of particles absorbed and p_a is the estimate of the probability of absorption, then the estimate of the standard deviation for the mean of n_a is given by

$$\text{StdDev}(\langle n_a \rangle) = n_t\sqrt{\frac{p_a(1-p_a)}{n_t}} = n_t\sqrt{\frac{\frac{n_a}{n_t}\left(1-\frac{n_a}{n_t}\right)}{n_t}} = \sqrt{\frac{n_a(n_t - n_a)}{n_t}} \tag{3.5}$$

The coding in Table 3.8P makes use of eqn 3.5.

For a binomial distribution, with a probability p for one of the possible events and a probability q = 1 - p for the other possible event, from eqn 3.3 the variance of p is given by:

$$\text{var}(p) = p(1-p) = pq = (1-q)q = \text{var}(q) \tag{3.6}$$

The equivalence between var(p) and var(q) follows because the error to which p can be estimated must be the same as the error to which 1 - p = q can be estimated. This result follows both from the logic of the binomial case, and from the fact that the variance of a variable plus a constant is equal to the variance of the variable. This equivalence between the variances of the alternative scores in a binomial distribution can be useful in some cases for checking the estimates of standard deviations obtained in Monte Carlo calculations. An example of this occurs in Chapter 9.

In Example 1.2 we estimate the value of π by using the rejection technique. Each point selected is either inside or outside a unit circle, and hence each score is either one or zero. Thus the sample forms a binomial distribution. The results of Example 1.2 are shown in Table 1.4P. To determine the number of scores of one that are produced in the calculation we could re-run the calculation and print out the counter 'npi' from Table 1.3P, or we could calculate this number using the result given in Table 1.4P. In the present notation this result is

$$\langle \pi \rangle = \frac{4n_l}{n_t}$$

Using the result in Table 1.4P for 10^6 samples we have $\langle \pi \rangle = 3.142096$ and thus

$$p = \frac{n_l}{n_t} = \frac{\langle \pi \rangle}{4} \approx 0.785524$$

Applying eqn 3.4 for the variance of a binomial distribution to this result, with $n_t = 10^6$, we see

$$\mathrm{var}\left(\frac{\langle \pi \rangle}{4}\right) = \frac{p(1-p)}{n_t} = 0.168476 \times 10^{-6}$$

Thus the standard deviation in the estimate of π for this case is given by

$$\sigma = 4\sqrt{\mathrm{var}\left(\frac{\langle \pi \rangle}{4}\right)} \approx 0.0016418$$

The standard deviation of the results of Example 1.2 can be calculated directly from eqn 3.4 by using the fact that the standard deviation is equal to the square root of the variance. A modification to the Fortran program shown in Table 1.3P that would incorporate this change is shown in Table 3.6. This routine produces the same results for both the estimates of π and the estimates of the standard deviations as the program of Table 1.3P, but does not include the variable 'sumsq.' The latter was used to save the square of the scores in order to apply eqn 1.9P.

Table 3.6. Fortran Program for Exercise 3.4

```
!  A Monte Carlo program to calculate Pi using rejection.  Points are selected
!  uniformly from a unit square.  Those points inside a unit circle are counted, others
!  rejected.  The area of the square is 1, that of the enclosed circle is pi/4.
!
     DOUBLE PRECISION fltrn, sumsq, pi, tries
     OPEN(unit=6,file='ranpi.txt')
     notries=1000000                         ! notries is the number of samples (tries)
     npi=0                                   ! initialize variables for scoring
     DO 100 i=1,notries
       x=fltrn()                             ! fltrn gives a random number (0,1)
       y=fltrn()                             ! selects a random pt (x,y) within square
       IF (x**2 + y**2 .LE. 1.0)THEN         ! score points inside circle
         npi = npi + 1                       ! add one, multiply by 4 later
       ENDIF
100  CONTINUE
     tries=FLOAT(notries); pi=FLOAT(npi)/tries
     stdev = SQRT((pi-pi**2)/tries)
     WRITE(6,14)notries,pi*4.0d0,stdev*4.0d0
 14  FORMAT (1x,'notries : ',i10,'   pi: ',f12.8,' std dev :',f12.8)
     STOP
     END
```

Chapter 4

Energy-Dependent Neutron Transport

Exercise 4.1

Statement of the problem

1. Repeat the calculation of Example 3.1 assuming isotropic scattering in the center of mass system using A = 1, but neglect energy changes. Compare your result with that obtained assuming isotropic scattering in the L system.

Solution

To solve this problem we start with the Fortran of Table 3.1P. However, instead of using the scattering model of Table 3.2P, which assumes isotropic scattering in the laboratory system, we allow for isotropic scattering in the center of mass by substituting subroutine 'Isocoll' of Table 4.5P for subroutine 'Isoout' of Table 3.2P. Subroutine 'Isocoll' does not have energy dependence, and therefore can be substituted directly for 'Isoout' provided we either modify the sequence of arguments in 'Isocoll' to omit the parameter 'A' or add the argument 'A' in the call to the scattering routine in Table 3.1P. Either way we must set the mass 'A' for the scatterer equal to one. The Fortran for a revised subroutine 'Isoout' is given in Table 4.1. This subroutine is based on the coding in Table 4.5P but the variable 'A' is removed from the argument list and defined internally. Combining the coding of this table with that of Table 3.1P allows us to solve the present exercise.

Running this code with 10^4 start particles produces the results shown in Table 4.2 for the number of collisions in the various spatial regions as a function of radius. The results are shown for the spatial intervals whose outer boundaries equal the indicated unit mfp values. In the table the current results are compared with the analogous results of Example 3.1, as shown in Table 3.3P. The current results for 0.1 and 10 mfp are somewhat less than, but comparable to, the earlier results. However, the number of scattering events at detector locations between these limits is significantly lower for center-of-mass isotropic scattering than for laboratory isotropic scattering.

Table 4.1. Revised Subroutine 'Isoout' for Exercise 4.1

```
      SUBROUTINE ISOOUT (U,V,W)
!     Subroutine ISOCOL1 of Table 4.5 of Primer.
!     Modified for Exercise 4.1. A is fixed as 1.
!
      DOUBLE PRECISION fltrn,PI,COSTH,PHI,UL,VL,WL,UF,VF,WF,SR,SINTH,T
      A=1.0
      PI=2.0d0*DACOS(0.0d0)
      COSTH = 2.D0*fltrn()-1.0D0  ! Cosine of random polar scatter angle selected
      PHI = fltrn()*2.D0*PI       ! Random azimuthal angle selected
      WL = (A*COSTH+1.D0)/DSQRT(A**2+2.D0*A*COSTH+1.D0) ! z-dir cosine,lab coordinates
      SINTH = DSQRT(1.D0-WL**2)   ! Azimuthal angle in lab coordinates
      UL = SINTH*DCOS(PHI)        ! direction cosine in lab coordinates
      VL = SINTH*DSIN(PHI)        ! direction cosine in lab coordinates
      IF(ABS(U).LT.0.9)THEN         ! select appropriate transformation
        SR=DSQRT(1.0D0-U*U)         ! Primary transformation T sub x
        UF=SR*UL+U*WL               ! final direction cosine U
        VF=-U*V*UL/SR+W*VL/SR+V*WL  ! final direction cosine V
        WF=-U*W*UL/SR-V*VL/SR+W*WL  ! final direction cosine W
    ELSE
        SR=DSQRT(1.0D0-V*V)         ! Alternate transformation T sub y
        UF=W*UL/SR-U*V*VL/SR+U*WL   ! final direction cosine U
        VF=SR*VL+V*WL               ! final direction cosine V
        WF=-U*UL/SR-V*W*VL/SR+W*WL  ! final direction cosine W
      ENDIF
      T=DSQRT(UF*UF+VF*VF+WF*WF)  ! renormalize to fix possible roundoff errors
      U=UF/T; V=VF/T; W=WF/T      ! set values for return
      RETURN
      END
```

Table 4.2. Selected Results for Exercise 4.1

Radius (mfp)	Example 3.1	Exercise 4.1
0.1	1114	1059
1	3055	1306
2	4812	1783
3	6386	2368
4	7712	2716
5	8205	3037
6	7954	3157
7	7265	2964
8	6127	2838
9	4473	2467
10	1808	1679

Discussion

In Example 3.1 we do not include estimates of the statistical uncertainties in the results. In general such estimates can provide an important indication of the level of confidence one can place in the results. The uncertainties could be calculated using eqns 2.43P and 2.44P. However, for the current problem it is possible to estimate the statistical uncertainties in the results using an approach similar to that used for estimating the uncertainties in the results for a binomial distribution, as derived in Exercise 3.4.

Let us define the quantity a such that the product ap, where p will be defined below, equals the probability that the first collision of a source particle will occur in a certain detector region. This product thus includes both the probability that a source particle will travel to the detector region without experiencing a collision, and the probability that the particle will experience a collision in the detector once it arrives. Let us also assume there is a probability c that a particle will emerge from the collision in the detector region, and that there is a probability b that a particle emerging from such a collision in the detector region will have a second collision in the same detector region. The quantity c is the non-absorption probability, as defined in section 3.3 of the *Primer*.

Let p equal the probability that a particle that has a collision in the detector region will have a second collision in the same detector region; i.e., p = cb. Since p is defined in terms of the second collision probability, the quantity a must provide the appropriate factor to ensure that the product ap is equal to the desired probability of a first collision in the detector region. In most problems a will be less than one. An exception would be when the intervening material is a strong absorber. In this case, as $c \rightarrow 0$, $a \rightarrow \infty$. Because p is the probability of a second collision in the detector given the first collision, 1−p is the probability there is not a second collision. We therefore have

ap	= probability of a first collision
$ap(1-p)$	= probability of one and only one collision
ap^2	= probability of a second collision
$ap^2(1-p)$	= probability of two and only two collisions
ap^3	= probability of three collisions
etc.	

Let S_1 be the expected number of collisions in the detector of interest. Then

$$S_1 = ap(1-p) + 2ap^2(1-p) + 3ap^3(1-p) + \cdots$$
$$= a\left[p + p^2 + p^3 + \cdots\right] \tag{4.1}$$

Using the equation for a geometric progression

$$\sum_{i=0}^{\infty} x^i = \frac{1}{1-x}, \ x < 1$$

allows us to express eqn 4.1 in closed form as

$$S_1 = \frac{ap}{1-p} \tag{4.2}$$

Let S_2 be the expected value of the sum of the squares of the numbers of collisions,

$$S_2 = ap(1-p) + 4ap^2(1-p) + 9ap^3(1-p) + \cdots$$
$$= ap + 3ap^2 + 5ap^3 + \cdots$$
$$= a\sum_{n=1}^{\infty}(2n-1)p^n$$

This is a power series that can be factored over its range of convergence to give

$$S_2 = ap\left[2\left(\sum_{n=0}^{\infty} p^n\right)^2 - \sum_{n=0}^{\infty} p^n\right]$$

Again using the closed-form expression for the sum of a geometric progression, we get

$$S_2 = ap\left[\frac{2}{(1-p)^2} - \frac{1}{1-p}\right] \tag{4.3}$$

From eqn 4.2 we see

$$p = \frac{S_1}{a + S_1} \tag{4.4}$$

and

$$1 - p = \frac{a}{a + S_1} \tag{4.5}$$

Applying these results to eqn 4.3 gives

$$S_2 = S_1 + \frac{2}{a} S_1^2 \tag{4.6}$$

Using eqn 2.29P with eqns 4.2 and 4.6, we obtain for the variance

$$\text{var} = S_2 - S_1^2 = S_1 + \left(\frac{2}{a} - 1\right) S_1^2 \tag{4.7}$$

The similarity between eqn 4.7 and eqn 3.3 for the variance of a binomial distribution is apparent.

A revised computer program for solving Example 3.1, which assumes isotropic scattering in the laboratory coordinates, that includes a calculation of the uncertainties in the results is shown in Table 4.3. This program uses the version of 'Isoout' shown in Table 3.2P. It calculates the standard deviation in the results using eqn 2.45P. It also calculates the values for the quantity a that would give the results of eqn 2.45P using the square root of eqn 4.7, and it calculates the standard deviations predicted by the square root of eqn 4.7 based on the assumption that a = 1. Selected results are shown in Table 4.4. All of the results shown here are based on 10^6 start particles in order to obtain accurate estimates for a. For comparison purposes, both the standard deviations that would be obtained by assuming a binomial distribution and those that would be obtained assuming counting statistics (the square root of the number of counts) are presented.

The values of the parameter a derived by matching the standard deviation results of eqn 2.45P with those from eqn 4.7 vary between about 0.62 and 1.14, with the largest value being obtained near the source and the smallest value occurring near the outer edge of the sphere. Results for the parameter at intermediate values of the radius reach a local maximum near 0.85 at five mfp, but for most of the volume of the sphere we find a ~ 0.8. Estimating the standard deviation from eqn 4.7 with a = 1 gives results that are low over most of the volume of the sphere, but only by about 10%. Thus the closed-form expression for the variance shown in eqn 4.7 with a = 1 appears to provide reasonable estimates of the uncertainty in the results for this example.

Table 4.3. Revised Program for Example 3.1 with Calculation of Standard Deviations

```
!       Example program for isotropic point source of neutrons
!       in a finite medium with isotropic scatter and no absorption
!         Score collision density flux in units of 0.1 mfp.
!         For convenience assume SIGMA = 1 cm**-1
!         U,V,W = particle Cartesian direction cosines
!         X,Y,Z = particle location Cartesian coordinates
!         D = particle flight path length upon leaving current collision
!         ncolsn(I) = number of collisions as a function of radius, 0.1 MFP resolution
      DOUBLE PRECISION rnd,fltrn
      DIMENSION ncolsn(100),ncolsq(100),ncolpart(100)
      OPEN(unit=8, file='output.txt', carriagecontrol='Fortran')
New_Case_Loop:  DO         ! The loop terminates if NS=0
      WRITE(*,'(1X,A\)')' Enter number of start particles '     ! Read input data
      READ(*,*)ns                                          ! number of start particles
      IF (ns.LE.0) EXIT New_Case_Loop                      ! set ns=0 to stop
      WRITE(*,'(1x,a\)')' seed '
      READ(*,*)iseed                      ! read seed for random number generator
      CALL rndin (iseed)                  ! set initial random number
      WRITE(8,'(1x,a\)')' seed ';  WRITE(8,10)iseed
   10 FORMAT(1x,I18)
      ncolsn=0.0                                   ! initialize ncolsn array to zero
      DO 300 n=1,ns                                ! Begin particle tracking
        u = 0.0; v = 0.0; w = 1.0                  ! set initial direction cosines
        xo = 0.0; yo = 0.0; zo = 0.0               ! set initial position
        ncolpart = 0
        DO 200
          rnd = fltrn()                            ! random number for picking distance
          d = -DLOG(rnd)                           ! distance to collision
          x = xo+u*d; y = yo+v*d; z = zo+w*d       ! coordinates of collision point
          r = SQRT(x**2+y**2+z**2)                 ! radius of collision point in sphere
!  It is not necessary to determine the type of reaction, problem involves scatter only
          i=INT(10.0d0*r)+1                        ! calculate mfp for bin number
          IF(i.GT.100)EXIT   ! exit DO 200 loop    ! if r>10 terminate particle
          ncolpart(i) = ncolpart(i)+1              ! score collision in bin i
          CALL ISOOUT(u,v,w)                       ! get new direction cosines
          xo = x; yo = y; zo = z                   ! update particle position
 200    CONTINUE
        DO i=1,100
          ncolsn(i)=ncolsn(i)+ncolpart(i)
          ncolsq(i)=ncolsq(i)+ncolpart(i)**2
        ENDDO
 300  CONTINUE
!     Problem complete.  Present results
      WRITE(8,14)ns
   14 FORMAT(' Number of start particles = ',I 10)
      fns=FLOAT(ns)
      DO i=1,100
        avgcol=ncolsn(i)/fns
        var=FLOAT(ncolsq(i))/fns-avgcol**2
        a=2.0*avgcol**2/(var-avgcol+avgcol**2)
        stdev=SQRT(var)
        WRITE (8,13)ncolsn(i),avgcol,stdev/SQRT(fns),a,stdev,SQRT(avgcol+avgcol**2)
      ENDDO
   13 FORMAT(I8,5e12.4)
END DO New_Case_Loop
      CALL RNDOUT(iseed)
      WRITE(8,'(1x,a\)')' end seed '
      WRITE(8,10)iseed                                 ending seed
      STOP;  END
```

Table 4.4. Selected Results for Example 3.1 with Standard Deviations, 10^6 particles

Radius	Expected number of collisions per start particle	Standard Dev eqn 2.45P	Standard Dev assuming a=1, eqn 4.7	Value of a to give eqn 2.45P result	Standard Dev assuming binomial distr	Standard Dev assuming counting statistics
0.1	0.1085	0.3426-3	0.3468-3	1.138	0.3110-3	0.3294-3
1	0.2893	0.6621-3	0.6107-3	0.7190	0.4534-3	0.5379-3
2	0.4873	0.9263-3	0.8513-3	0.7806	0.4998-3	0.6981-3
3	0.6442	0.1114-2	0.1029-2	0.8200	0.4788-3	0.8026-3
4	0.7469	0.1232-2	0.1142-2	0.8394	0.4348-3	0.8642-3
5	0.7974	0.1290-2	0.1197-2	0.8458	0.4019-3	0.8930-3
6	0.7937	0.1287-2	0.1193-2	0.8444	0.4046-3	0.8909-3
7	0.7325	0.1220-2	0.1127-2	0.8305	0.4627-3	0.8559-3
8	0.6145	0.1082-2	0.9961-3	0.8095	0.4867-3	0.7839-3
9	0.4396	0.8698-3	0.7955-3	0.7575	0.4963-3	0.6630-3
10	0.1839	0.5094-3	0.4666-3	0.6183	0.3874-3	0.4288-3

A similar comparison can be made for the problem with center-of-mass scattering. Running the program in Table 4.3, along with the 'Isoout' routine of Table 4.1, gives the results shown in Table 4.5. As in Table 4.4 the standard deviation results calculated using eqn 2.45P, and those obtained using eqn 4.7 with $a = 1$, are shown. Values of a for which these two results would be equal are also shown. In this case the values of the parameter a vary between about 0.69 and 1.7, and for most of the volume of the sphere the value of a is near 0.7. Again the use of $a = 1$ gives error estimates that are generally smaller than those obtained from eqn 2.45P but are within 10% of the latter values.

Table 4.5. Selected Results for Exercise 4.1 with Standard Deviations, 10^6 particles

Radius	Expected number of collisions per start particle	Standard Dev eqn 2.45P	Standard Dev assuming a=1, eqn 4.7	Value of 'a' to give eqn 2.45P result	Standard Dev assuming binomial distr	Standard Dev assuming counting statistics
0.1	0.1007	0.3199-3	0.3328-3	1.715	0.3009-3	0.3173-3
1	0.1311	0.3914-3	0.3851-3	0.8762	0.3375-3	0.3621-3
2	0.1803	0.4883-3	0.4613-3	0.7177	0.3844-3	0.4246-3
3	0.2297	0.5728-3	0.5314-3	0.6977	0.4206-3	0.4793-3
4	0.2714	0.6387-3	0.5874-3	0.7007	0.4447-3	0.5210-3
5	0.2970	0.6763-3	0.6207-3	0.7097	0.4569-3	0.5450-3
6	0.3106	0.6976-3	0.6380-3	0.7079	0.4627-3	0.5573-3
7	0.3050	0.6898-3	0.6309-3	0.7051	0.4604-3	0.5523-3
8	0.2850	0.6608-3	0.6051-3	0.6971	0.4514-3	0.5339-3
9	0.2453	0.6007-3	0.5527-3	0.6853	0.4303-3	0.4953-3
10	0.1688	0.4642-3	0.4441-3	0.7575	0.3746-3	0.4109-3

The results of this exercise along with the results of Example 3.1 are plotted as a function of radius in Figure 4.1. The labels L and C designate the results for isotropic scattering in the laboratory and center-of-mass coordinate systems, respectively. For hydrogen, isotropic neutron scattering in the center of mass results in significant forward peaking in the post-collision laboratory angular distribution. As a result the number of collisions experienced by a neutron before escaping from the geometry is greatly reduced compared with that for a calculation that assumes isotropic scattering in the laboratory system. That is, there is much less backscatter in the present exercise than in the earlier example that assumed isotropic scattering in the laboratory system. A simple computation predicts that a source neutron will experience about 57 collisions on the average before it escapes from the system in the laboratory isotropic scattering case versus about 24 collisions in the center-of-mass isotropic scattering case. As a consequence the flux in the interior of the sphere, but far from the source, is significantly reduced in the current calculation compared with the laboratory isotropic scattering calculation.

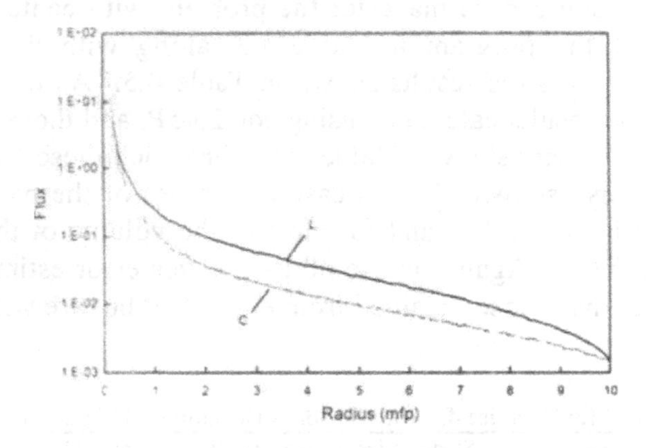

Figure 4.1. Results for Exercise 4.1

For the present problem the flux is not strongly dependent on the scattering model near the source and the outer boundary, and the results of this exercise are close to those of Example 3.1 at these points. Near the source the total flux is dominated by uncollided neutrons and thus is independent of the type of scattering that occurs in the bulk material. At such points we find the parameter a of eqn 4.7 to be greater than one. Because there is no absorption in the problem, all source particles eventually escape from the geometry. Thus at the outer boundary the fluxes are determined primarily by the radius and the number of escaping particles. With no absorption these are independent of the type of scattering assumed in the bulk material. Therefore, the fluxes at the outer boundary are similar for isotropic scattering in the center-of-mass system and isotropic scattering in the laboratory system.

The derivation of eqn 4.7 applies to analog Monte Carlo transport. Although it is not needed for this exercise, this result can be extended to the case in which the technique called "survival biasing" is used. Survival biasing is discussed in Chapter 6 of the *Primer*. When this technique is used the "weight" of the particle being tracked is reduced after each collision by multiplying the incoming weight by the non-absorption probability. The weight emerging from the collision thus represents the fraction of the particle that survives.

To determine the expected variance of a volumetric detector tally in a calculation that includes survival biasing, let us again define the product ap as the probability of a first collision in the detector region. In this case, in addition to including the factors required in the analog case, the quantity a must account for the reduced weight of the particle when it undergoes subsequent collisions in the detector. As before we assume c equals the probability that a particle will emerge from the collision, and b equals the probability of a particle that emerges from a collision in the detector region having a second collision in the same detector region. Again we define p = cb; however, the total probability that a particle having a first collision in the detector region will have a second collision in the same detector region is now equal to b rather than p because survival biasing reduces the weight of the surviving particle rather than reducing the probability that the particle will survive the collision.

To continue this derivation with survival biasing it is useful to represent the probability of a first collision as acb, which follows from the definition of p. Therefore

acb = probability of a first collision
$acb(1-b)$ = probability of one and only one collision
acb^2 = probability of a second collision
$acb^2(1-b)$ = probability of two and only two collisions
acb^3 = probability of three collisions
etc.

As before let S_1 be the expected number of collisions in the detector such that

$$S_1 = acb(1-b) + (1+c)acb^2(1-b) + \left(1+c+c^2\right)acb^3(1-b) + \left(1+c+c^2+c^3\right)acb^4(1-b) + \cdots$$

This can be written as

$$S_1 = acb - acb^2 + (1+c)(acb^2 - acb^3) + (1+c+c^2)(acb^3 - acb^4) + \cdots$$
$$= a\left[cb + (cb)^2 + (cb)^3 + \cdots\right]$$

Since p = cb, this can also be written as

$$S_1 = a\left(p + p^2 + p^3 + \cdots\right) = \frac{ap}{1-p} \tag{4.8}$$

This result is the same as that for the analog scattering case, eqn 4.2, since survival biasing is a fair game and does not change the expected value of the desired result.

Let S_2 be the expected value of the sum of the squares of the scores at collisions for the case with survival biasing. Then

$$S_2 = acb(1-b) + (1+c)^2 acb^2(1-b) + \left(1+c+c^2\right)^2 acb^3 + \cdots$$

This can be rewritten as

$$S_2 = \left(acb - acb^2\right) + (1+c)^2\left(acb^2 - acb^3\right) + \left(1+c+c^2\right)\left(acb^3 - acb^4\right) + \cdots$$

Canceling like terms gives

$$S_2 = acb + (2+c)a(cb)^2 + \left(2+2c+c^2\right)a(cb)^3 + \left(2+2c+2c^2+c^3\right)a(cb)^4 + \cdots$$
$$= ap + (2+c)ap^2 + \left(2+2c+c^2\right)ap^3 + \left(2+2c+2c^2+c^3\right)ap^4 + \cdots$$

This is a power series that can be factored over its range of convergence to give

$$S_2 = ap\left[2\sum_{n=0}^{\infty}(cp)^n\sum_{n=0}^{\infty}p^n - \sum_{n=0}^{\infty}(cp)^n\right]$$

Using the closed-form expression for the sum of a geometric progression gives

$$S_2 = ap\left[\frac{2}{(1-p)(1-cp)} - \frac{1}{1-cp}\right]$$

As before we have eqns 4.4 and 4.5, which follow from eqn 4.8, and we then obtain

$$S_2 = \frac{aS_1 + 2S_1^2}{a + S_1 - cS_1} \qquad (4.9)$$

Thus, using eqn 2.29P along with eqns 4.8 and 4.9, the variance for the survival biasing case is given by

$$\text{var} = \left(\frac{a}{a + S_1 - cS_1}\right)S_1 + \left[\left(\frac{2}{a + S_1 - cS_1}\right) - 1\right]S_1^2 \qquad (4.10)$$

When c =1 the particles survive all collisions with a weight of one and eqn 4.10 reduces to the analog case of eqn 4.7.

For a simple comparison to determine the effectiveness of survival biasing, consider the case where a = 1, S_1 = 1 and c = 0.5. With survival biasing (eqn 4.10) the variance is one while without survival biasing (eqn 4.7) the variance is two. Because the variance is inversely proportional to the number of particles tracked, in this case the efficiency of survival biasing depends on finding a scheme that implements the technique without increasing the computer run time by more than a factor of two. The efficiency of variance reduction techniques, such as survival biasing, is the subject of Chapter 6 of the *Primer*.

Exercise 4.2

Statement of the problem

2. Estimate the age of neutrons at the Cd cutoff from a fission source in Be (A = 9). For the total microscopic cross section of Be use σ = 6 barns for E < 50 keV and use 14.782 - 1.8687 $\log_{10}(E)$ for E > 50 keV where E is in eV. The density of Be is 1.85 gms per cm^3. Assume c = 1.

a. Assume isotropic elastic scatter in the center of mass system.

b. Include inelastic scattering assuming the target nuclei have an internal excitation energy level of 100 keV, and that the probability of inelastic scatter above this energy is 0.5. Assume inelastic scattering is isotropic in the laboratory system and that the energy loss experienced by the neutron in inelastic scattering events is the excitation energy; i.e., $E_f = E_o - 10^5$ eV.

Solution Part a

This problem is similar to Example 4.3. As in that example we calculate $\langle r^2 \rangle$ and estimate τ using eqn 4.52P. The number density of Be atoms at the stated density is $N_{Be} = N_0\rho/A$, where N_o is Avagadro's number, ρ is the density, and A is the atomic mass. Thus, for a nominal beryllium density of 1.85 g/cm^3, we have $N_{Be} \approx 0.1238$ atoms/barn-cm. With this result we can adapt the Fortran in Tables 4.7P through 4.10P to solve the current problem. Specifically, the Fortran of Table 4.7P is replaced with that of Table 4.6. Subroutine 'Input,' given in Table 4.8P, is not needed in this calculation. Subroutine 'Isocol,' given in Table 4.9P, is used with no change, and is not repeated here. Table 4.10P, which shows the subroutines used for calculating the cross sections, is replaced by the coding in Table 4.7. In the latter table we make use of the fact that $\log_{10}x = \log_e x/\log_e 10$. In the coding of Table 4.6 we have set the outer radius of the system to 100 cm and the Cd cutoff energy to 0.415 eV. As in Example 4.3 we tally particles that escape from the system and we terminate the tracking of particles when they reach the cutoff energy. By tallying the number of neutrons that escape before reaching the cutoff energy we can determine whether the system radius we have chosen is adequate. If few or no neutrons escape before reaching the cutoff energy we can be sure that we have not introduced systematic bias into our result. Running this code with 10^4 start particles gives $\langle r^2 \rangle = 367.1 \pm 3.3$ cm^2, and thus $\tau \approx 61.18$ cm^2.

Solution Part b

In order to include inelastic scattering in the above coding we must modify the treatment of the scattering by testing whether the neutron energy is above the inelastic threshold, which in this case is assumed to be 100 keV, and if so treat the scattering as inelastic with a probability of 0.5. If the scattering is inelastic we call 'Isoout' to obtain a post-collision direction of travel that is isotropic in the laboratory coordinates. If the scattering is not inelastic it is assumed to be elastic and isotropic in the center of mass. When the energy of the neutron is less than or equal to 100 keV we assume inelastic scattering is not possible and that all scattering is elastic and isotropic in the center of mass.

Table 4.6. Main Program for Exercise 4.2, Part a

```
!      Scattering is energy-dependent, elastic, isotropic in center of mass.
!      For Be (1.85g/cc), set bedens=0.1238
!      A Watt fission spectrum is used for source neutrons
!      ef = Cd cutoff energy;    e = Energy of particle after collision
!      U,V,W =  Direction cosines,      X,Y,Z = Cartesian coordinates of particle
!      NMAX = Number of start particles
!      100 cm sphere
       double precision fltrn
       DATA PI/3.14159265/
       DATA bedens/0.1238/,ef/0.415/,radius/100/,a/9./
Loop_Over_Problem: DO
       WRITE(*,*)                                          ! Read input
       WRITE(*,'(1X,A\)')' Enter number of start particles '
       READ(*,*)NMAX; IF(NMAX.LE.0)STOP                    ! Number of start particles
       WRITE(*,'(1X,A\)')' Enter start random number seed '
       READ(*,*)ISEED
       CALL rndin(iseed)                                   ! set starting random number
       nr=0; nl=0; SCOR=0.; r2bar=0.; r2barsq=0.; N=0      !  initialize for scoring
  Loop_Over_Particles: DO N = 1, NMAX
       U = 0.; V = 0.; W = 1.                        !  Source direction cosines
       X = 0.; Y = 0.; Z = 0.                        !  Source location
  65   xtest=-dlog(fltrn());ytest=-dlog(fltrn())      !  pick a point
       IF((ytest-xtest-1.)**2.gt.4.*xtest) GOTO 65    !  reject if fails test
       e = 2.*xtest*1.e6                             !  value selected from fission spectrum
       DO WHILE (e.GT.ef)                            !  while above cutoff energy
         CALL Be(bexs,e,bedens)                      ! cross sections at energy E
         d = -dlog(fltrn())/bexs                      !  Get a flight path D in cm
         X = X+D*U; Y = Y+D*V; Z = Z+D*W              !  collision site
         r2 = x**2+ y**2+z**2;    r = sqrt(r2)        !  radius of collision site
         IF (r.GE.radius) THEN                     !  if still in sphere, go to isocol
           nl = nl + 1                                 !  particle escaped (leaked)
           CYCLE Loop_Over_Particles                   !  get next particle
         ENDIF
         CALL ISOCOL (u,v,w,A,E)                      !  get new direction and energy
       END DO
       r2bar=r2bar+r2; r2barsq=r2barsq+r2*r2         !  below cutoff energy, score r**2
  END DO Loop_Over_Particles                         !  loop to next particle
       r2bar=r2bar/FLOAT(nmax); r2barsq=r2barsq/FLOAT(nmax) !  problem complete
       stdev = SQRT((r2barsq - r2bar*r2bar)/FLOAT(nmax))
       WRITE(*,27)r2bar,stdev,r2bar/6.0
  27   FORMAT(' Mean sq dist to abs ',1p2e12.3,' age ',1pe12.3)
       WRITE (*,13)NL
  13   FORMAT(' Particles leaking above ef =',I6)
END DO Loop_Over_Problem
       END
```

Table 4.7. Subroutine 'Be' for Exercise 4.2, Part a

```
       SUBROUTINE Be(bexs,e,bedens)
       bexs=6.
       IF(e.GT.50000.)bexs=14.782 - 1.8687*ALOG(e)/ALOG(10.)
       bexs=bexs*bedens
       RETURN
       END
```

The Fortran in Table 4.8 accommodates this change. The subroutine 'Isoout' used here is the same as that used in Example 3.1, Table 3.2P, and is not repeated. Subroutine 'Isocol,' given in Table 4.9P, is also included without change. The routines in Table 4.10P, which are used for calculating the cross sections in Example 4.3, are again replaced by the subroutine in Table 4.7.

Table 4.8. Main Program for Exercise 4.2, Part b

```
!                          PROGRAM for Exercise4.2b
!      Scattering is energy-dependent; elastic is isotropic in center of mass,
!      inelastic is isotropic in lab.  For Be (1.85g/cc), set bedens=0.1238
!      A Watt fission spectrum is used for source neutrons
!      ef = Cd cutoff energy;    e = Energy of particle after collision
!      U,V,W =  Direction cosines,      X,Y,Z = Cartesian coordinates of particle
!      NMAX = Number of start particles
!      100 cm sphere
       double precision fltrn
       DATA PI/3.14159265/
       DATA bedens/0.1238/,ef/0.415/,radius/100/,a/9./
       DATA enon/1.E5/
Loop_Over_Problem: DO
       WRITE(*,*)                                          ! Read input
       WRITE(*,'(1X,A\)')' Enter number of start particles '
       READ(*,*)NMAX; IF(NMAX.LE.0)STOP                    ! Number of start particles
       WRITE(*,'(1X,A\)')' Enter start random number seed '
       READ(*,*)ISEED
       CALL rndin(iseed)                                       ! set starting random number
       nr=0; nl=0; SCOR=0.; r2bar=0.; r2barsq=0.; N=0          ! initialize for scoring
  Loop_Over_Particles: DO N = 1, NMAX
       U = 0.; V = 0.; W = 1.                        ! Source direction cosines
       X = 0.; Y = 0.; Z = 0.                        ! Source location
   65  xtest=-dlog(fltrn());ytest=-dlog(fltrn())       ! pick a point
       IF((ytest-xtest-1.)**2.gt.4.*xtest) GOTO 65    ! reject if fails test
       e = 2.*xtest*1.e6                             ! value selected from fission spectrum
       DO WHILE (e.GT.ef)                            ! while above cutoff energy
         CALL Be(bexs,e,bedens)                      ! cross sections at energy E
         d = -dlog(fltrn())/bexs                       ! Get a flight path D in cm
         X = X+D*U; Y = Y+D*V; Z = Z+D*W               ! collision site
         r2 = x**2+ y**2+z**2;      r = sqrt(r2)       ! radius of collision site
         IF (r.GE.radius) THEN                         ! if still in sphere, go to 12
           nl = nl + 1                                 ! particle escaped (leaked)
           CYCLE Loop_Over_Particles                   ! get next particle
         ENDIF
         IF(e.GT.enon.and.fltrn().gt.0.5)THEN         ! test for inelastic threshold
           e=e-enon                                   ! inelastic scattering
           CALL ISOOUT(u,v,w)
         ELSE
           CALL ISOCOL (u,v,w,A,E)                     ! elastic scattering
         ENDIF
      END DO
       r2bar=r2bar+r2; r2barsq=r2barsq+r2*r2         ! below cutoff energy, score r**2
  END DO Loop_Over_Particles                         ! loop to next particle
       r2bar=r2bar/FLOAT(nmax); r2barsq=r2barsq/FLOAT(nmax) !  problem complete
       stdev = SQRT((r2barsq - r2bar*r2bar)/FLOAT(nmax))
       WRITE(*,27)r2bar,stdev,r2bar/6.0
   27  FORMAT(' Mean sq dist to abs ',1p2e12.3,' age ',1pe12.3)
       WRITE (*,13)NL
   13  FORMAT(' Particles leaking above ef =',I6)
END DO Loop_Over_Problem
       END
```

Running this coding with 10^4 start particles gives $\langle r^2 \rangle = 358.8 \pm 3.4$ cm^2, and thus $\tau \approx 59.8$ cm^2. The measured value of the Fermi age of fission neutrons in Be to thermal energy is 85 cm^2. Although the Fermi age to the

Cd cutoff energy should be somewhat less than 85 cm^2, because the cross sections used in this exercise are but simple approximations to the actual values neither part of this exercise gives the correct age at Cd cut-off for fission neutrons in Be.

Exercise 4.3

Statement of the problem

3. Consider the slowing down of fast neutrons in a nonabsorbing material that can undergo both elastic and inelastic scatter. Assume the elastic scattering is isotropic in the center of mass system, and the inelastic scattering is isotropic in the laboratory system. Further assume the material has a single internal excitation energy level of ε = 200 keV, and assume the inelastic scatter probability is 0.75 for neutrons of energy $E > \varepsilon$. Assuming $A = 12$, determine the mean number of collisions required to reduce the energy of a neutron from 1 MeV to the Cd cutoff in this material. Assume the energy loss during inelastic scattering is the excitation energy.

Solution

This problem combines the features of Example 4.1 and Exercise 4.2b. We therefore use portions of the Fortran associated with both of these problems to solve this exercise. Because we do not need to track the particles through the geometry in order to obtain a solution, we do not need to consider the fact that the inelastic scattering is isotropic in the laboratory system; however, to determine the correct energy change in elastic scattering events we do need to consider the fact that such scattering is assumed to be isotropic in the center-of-mass system.

For this problem we use the main program shown in Table 4.9. This program is similar to that of Example 4.1, shown in Table 4.1P, but it tracks the neutron energy in units of eV and it uses double precision for the energy calculations. To incorporate inelastic scattering we modify the version of subroutine 'Isocol2' shown in Table 4.2P. This revised routine, shown in Table 4.10, subtracts the excitation energy from the particle energy with probability 0.75 if the particle energy is greater than the excitation energy. If the particle energy is less than or equal to the excitation energy, or if the scattering is not inelastic, the routine calculates the post-collision energy using the method of Table 4.2P.

Running the program of Tables 4.9 and 4.10 for 10^4 start particles and a start energy of 10^6 eV we obtain the result 84.97 ± 0.07 for the mean number of collisions required to reduce the energy of a neutron from 1.0 MeV to Cd cutoff for A = 12. As in Example 4.1 we assume the Cd cutoff energy to be 0.415 eV.

Table 4.9. Main Program for Exercise 4.3

```
!       Determine the mean number of collisions to Cd cutoff (0.415 eV)
!       for a target of atomic mass A, neutrons with start energy E (eV)
        REAL*8 E,EP,A,ECD,EO
        DATA ECD/0.415d0/
    4   WRITE (*,'(1X,A\)') ' Enter target mass A, set=0 to stop:  '    ! Input variables
        READ (*,*) A;       IF (A.LE.0.1) STOP                            !
        WRITE (*,'(1X,A\)') ' Enter neutron energy E (eV):  '            !
        READ (*,*) EO;      IF(EO.LT.ECD) STOP                            !
        WRITE (*,'(1X,A\)') ' Enter number of start particles N:  '      !
        READ (*,*) N;       IF(n.LE.0) STOP                               !
        WRITE (*,'(1X,A\)') ' Enter random number seed (an integer):  '!
        READ (*,*) ISEED;  CALL rndin(iseed)                            ! Input complete
        TALLY = 0.; TALLYSQ = 0.                      ! set tally variables to zero
Loop_Over_Particles:DO I = 1,N          ! 'Track' particles, loop over N start particles
    EP = EO                             ! start each particle with energy EO
    NSCAT = 0                           ! NSCAT is the number of scatterings
  Loop_Over_Energy:DO                   ! loop while above cutoff energy
        IF (EP.LT.ECD) EXIT Loop_Over_Energy  ! score if below cutoff energy
        E = EP                          ! update pre-collision energy
        CALL ISOCOL2(E,EP,A)            ! get post collision energy EP
        NSCAT = NSCAT + 1               !
        IF(NSCAT.LT.10000)CYCLE Loop_Over_Energy
!
!       (If 10000 scatterings do not suffice to reach the Cd cutoff, the
!       downscatter is terminated.  User set the energy very high or the target
!       mass very large so the problem is unrealistic.)
!
        WRITE (*,*) ' More than 10000 collisions required, particle ',I
        CYCLE Loop_Over_Particles
  END DO Loop_Over_Energy
    TALLY = TALLY + NSCAT;  TALLYSQ = TALLYSQ + NSCAT**2    ! below cutoff, tally
particle
END DO Loop_Over_Particles
!       Start particle loop complete, determine mean, standard deviation
!       and fractional standard deviation of mean number of collisions
!
        XCOL = TALLY/N;       XCOL2 = TALLYSQ/N
        StDev = SQRT((XCOL2-XCOL**2)/N);  FSD=StDev/XCOL
        WRITE(*,*)' Mean number of collisions to Cd cutoff = ',XCOL
        WRITE(*,*)' Standard deviation = ',StDev
        WRITE(*,*)' Fractional standard deviation = ',FSD
        GO TO 4
        END
```

Discussion

In the absence of inelastic scattering we can use eqns 4.17P, 4.18P and 4.21P to estimate the number of collisions required to reduce a 1-MeV neutron to the Cd cutoff energy for A = 12. For this case we find $\alpha = 0.716$, which predicts a logarithmic energy decrement of $\zeta = 0.1578$. This result gives $N_L = 95.94$ and $N_G = 93.14$ for the linear and geometric average number of collisions required to reduce the neutron energy from 1 MeV to

0.415 eV, respectively. Because the energy loss from inelastic scattering is a significant factor in this exercise our calculated result of about 85 collisions is less than either of the analytic elastic scattering results.

Table 4.10. Revised Subroutine 'Isocol2' for Exercise 4.3

```
      SUBROUTINE ISOCOL2 (E,EP,A)
!
!     Subroutine ISOCOL2 calculates the result of isotropic collisions
!     of neutrons of energy E with nucleii of mass A.  The post-collision
!     energy EP in the laboratory system is determined assuming
!     isotropic scattering in the center of mass system.
!
      REAL*8 fltrn,EIN,E,EP,A,COSTH
      DATA EIN/200000.0D0/
      IF(E.LE.EIN.OR.fltrn().GT.0.75)THEN
        COSTH = 2.*fltrn()-1.                    ! cosine of random polar angle
        EP = E*(A*A+2.*A*COSTH+1.)/(A+1.)**2     ! Outgoing energy EP
      ELSE
        EP=E-EIN                                 ! Outgoing energy for inelastic scat
      ENDIF
      RETURN
      END
```

This exercise defines a discrete source energy and, therefore, the result has a discontinuity at any source energy that is an integral multiple of the inelastic threshold of 0.2 MeV. For start particles with energies E_0 in the range from 1.0 MeV to slightly above 0.8 MeV the result will be continuously decreasing with decreasing E_0, beginning at a value of about 85 collisions for $E_0 = 1$ MeV. For start particles with E_0 from 1.2 MeV to slightly above 1.0 MeV the result will again decrease continuously with decreasing energy, ending at a value of about 66 collisions for $E_0 = 1 + \delta$ MeV, where δ is a small, positive number.

This discontinuity, which occurs exactly at the required source energy, may cause the Fortran used to solve the exercise to produce an interesting and possibly unexpected anomaly. Because of the manner in which digital computers treat floating point numbers, and specifically the way in which numbers less than one are represented, with some compilers this code may give an incorrect result. For example, let us repeat the above calculation while following the neutron energy in units of MeV instead of eV. To do this we revise the program of Table 4.9, as shown in Table 4.11. The revised, MeV-version of subroutine 'Isocol2' is shown in Table 4.12.

Executing the revised program with a start particle energy of 1.0 MeV, and again using 10^4 start particles, we obtain the incorrect result of 65.72 ± 0.34 collisions. This error results from the way the energy is represented in the computer. After four consecutive inelastic scattering events the computer assigns an energy to the neutron that is slightly greater than the inelastic threshold. As a result, according to the statement of the problem a subsequent inelastic scattering reaction can then occur. After this fifth inelastic scattering event, which will occur for 75% of such particles, the energy of the particle falls below the cut-off value. Thus about 23.73% of the particle tracks are incorrectly stopped after five scattering events. Using this result for the number of source particles that incorrectly undergo only five scattering events to reach the Cd cut-off, along with the correct result of 84.97 for the remaining 76.27% of the source particles, we obtain an answer of about 66 as the mean number of collisions required to slow the source neutrons to the Cd cut-off. This result is close to the incorrect result produced by the Fortran program when units of MeV are used.

Table 4.11. Revised Main Program for Exercise 4.3, Energy in MeV

```
!       Determine the mean number of collisions to Cd cutoff (0.415E-6 MeV)
!       for a target of atomic mass A, neutrons with start energy E (MeV)
      REAL*8 E,EP,A,ECD,EO
      DATA ECD/0.415d-6/
    4 WRITE (*,'(1X,A\)') ' Enter target mass A, set=0 to stop:  '   ! Input variables
      READ (*,*) A;       IF (A.LE.0.1) STOP                          !
      WRITE (*,'(1X,A\)') ' Enter neutron energy E (MeV):  '           !
      READ (*,*) EO;      IF(EO.LT.ECD) STOP                           !
      WRITE (*,'(1X,A\)') ' Enter number of start particles N:  '     !
      READ (*,*) N;       IF(n.LE.0) STOP                              !
      WRITE (*,'(1X,A\)') ' Enter random number seed (an integer):  '!
      READ (*,*) ISEED;  CALL rndin(iseed)                             ! Input complete
      TALLY = 0.; TALLYSQ = 0.                      ! set tally variables to zero
Loop_Over_Particles:DO I = 1,N              ! 'Track' particles, loop over N start particles
    EP = EO                                 ! start each particle with energy EO
    NSCAT = 0                               ! NSCAT is the number of scatterings
  Loop_Over_Energy:DO                       ! loop while above cutoff energy
      IF (EP.LT.ECD) EXIT Loop_Over_Energy  ! score if below cutoff energy
      E = EP                                ! update pre-collision energy
      CALL ISOCOL2(E,EP,A)                  ! get post collision energy EP
      NSCAT = NSCAT + 1                     !
      IF(NSCAT.LT.10000)CYCLE Loop_Over_Energy
!
!       (If 10000 scatterings do not suffice to reach the Cd cutoff, the
!       downscatter is terminated.  User set the energy very high or the target
!       mass very large so the problem is unrealistic.
!
      WRITE (*,*) ' More than 10000 collisions required, particle ',I
      CYCLE Loop_Over_Particles
  END DO Loop_Over_Energy
    TALLY = TALLY + NSCAT;  TALLYSQ = TALLYSQ + NSCAT**2    ! below cutoff, tally
particle
END DO Loop_Over_Particles
!       Start particle loop complete, determine mean, standard deviation
!       and fractional standard deviation of mean number of collisions
!
      XCOL = TALLY/N;          XCOL2 = TALLYSQ/N
      StDev = SQRT((XCOL2-XCOL**2)/N);  FSD=StDev/XCOL
      WRITE(*,*)' Mean number of collisions to Cd cutoff = ',XCOL
      WRITE(*,*)' Standard deviation = ',StDev
      WRITE(*,*)' Fractional standard deviation = ',FSD
      GO TO 4
      END
```

Although in a "real" application it is unlikely that the anomaly evinced in this exercise would occur, this problem illustrates one way in which a user may obtain incorrect answers to a Monte Carlo problem even when the computer program used to solve the problem is apparently correct. The dimensional units in which one solves a problem should be irrelevant. However, digital computers can process numbers in different ways depending on the magnitude of the numbers. Given the parameters of this exercise, just over 30% of the 1-MeV source particles will have four consecutive inelastic scattering reactions and thus reach an energy of exactly 0.2 MeV. According to the statement of the problem the inelastic reaction is not possible for these particles. That is, since their energy is exactly equal to the threshold for inelastic scattering, and the particle energy must exceed the threshold value to undergo inelastic scattering, such particles cannot undergo inelastic scattering. Therefore the correct answer is as stated above.

Table 4.12. Revised Subroutine 'Isocol2' for Exercise 4.3, Energy in MeV

```
      SUBROUTINE ISOCOL2 (E,EP,A)
!
!     Subroutine ISOCOL2 calculates the result of isotropic collisions
!     of neutrons of energy E with nucleii of mass A.  The post-collision
!     energy EP in the laboratory system is determined assuming
!     isotropic scattering in the center of mass system.
!
      REAL*8 fltrn,EIN,E,EP,A,COSTH
      DATA EIN/0.20D0/
      IF(E.LE.EIN.OR.fltrn().GT.0.75)THEN
        COSTH = 2.*fltrn()-1.                  ! cosine of random polar angle
        EP = E*(A*A+2.*A*COSTH+1.)/(A+1.)**2   ! Outgoing energy EP
      ELSE
        EP=E-EIN                               ! Outgoing energy for inelastic scat
      ENDIF
      RETURN
      END
```

Chapter 5

A Probabilistic Framework Code

Exercise 5.1

Statement of the problem

1. Repeat Example 5.1 using a sphere with radius 15 mfp. Compare the Monte Carlo results to the diffusion theory solution (eqn 3.34P.)

Solution

As in the examples in the *Primer*, the exercises in Chapters 5 through 9 are solved using the Probabilistic Framework Code (PFC) unless otherwise stated. There are two straightforward ways to modify the coding developed for solving Example 5.1 to model a 15-mfp sphere and thereby solve this exercise. Because the geometry in Example 5.1 is that of a sphere with a radius of 10 cm, one method is to change the cross section given in Table 5.24P. If the total cross section is changed from 1.0 cm^{-1} to 1.5 cm^{-1} then the radius of 10 cm changes from 10 mfp to 15 mfp. With this change the subroutines of Example 5.1 can be used without modification to calculate the flux in a 15-mfp sphere. Since the spatial resolution required for the results is not specified in the statement of the exercise, one may choose to use the resulting 0.15-mfp spatial resolution rather than the 0.1-mfp resolution used in the example.

The second method that one might use to solve this exercise is to leave the cross section at 1.0 cm^{-1} but change the geometry description so that the sphere has a radius of 15 cm. However, in the absence of other modifications the spatial distribution of the scoring bins for the flux will still be based on the 10-cm geometry and thus will not be uniformly distributed as a function of radius. To fix this the user must modify the radial intervals used to score collisions to reflect the new geometry. This can be done by modifying subroutine 'Col,' as given in Table 5.25P, so that intervals of 0.15 mfp are used for scoring purposes. Again, if the user chooses to accept a spatial resolution of 0.15 mfp, the problem can be correctly solved by making these changes and leaving the total cross section at 1 cm^{-1}.

If, on the other hand, the user wishes to maintain intervals of 0.1 mfp for the spatial resolution of the flux, the dimensions of the arrays in common 'stat' that define the scoring intervals (used in subroutines 'Col' and 'Stats') and the 'DO' loops in subroutine 'Stats' (Table 5.26P) need to be changed. That is, in Example 5.1 the dimensions of the scoring arrays in common 'stat' are set to 100. These dimensions need to be increased to 150 if uniform intervals of 0.1 mfp are to be used in the new calculation. Likewise the limits of the 'DO' loops are set to accommodate the 100 spatial intervals of Example 5.1. These need to be changed to 150 intervals. Obviously these changes in the scoring arrays, along with changes in the widths of the intervals, can be used in combination with an increase in the total cross section to solve the problem with uniform 0.1-mfp spatial resolution while retaining a 10-cm outer radius. Alternatively, the outer radius can be increased to 15 cm and the total cross section left unchanged.

The latter method of solution of the exercise, with a spatial resolution of 0.1 mfp, is illustrated here. The list of subroutines modified for this solution is given in Table 5.1. The new subroutines 'Col' and 'Stats' are shown in Tables 5.2 and 5.3. The scoring arrays in 'Col' have been re-dimensioned as indicated above while, in addition to re-dimensioning the scoring arrays, 'Stats' has been modified to write only the volume-normalized Monte Carlo results on the output file.

Table 5.1. Modified Subroutines for Exercise 5.1

Subroutine	Location
'Col'	Table 5.2
'Stats'	Table 5.3

The monoenergetic diffusion theory solution for the particle flux at a distance r from a point source in an isotropic, pure scattering material is given by eqn 3.34P. Since the flux estimates from the present Monte Carlo calculation are based on collisions within spherical shells, the tallies in subroutine 'Col' provide volume-averaged flux results. To compare our calculated fluxes with the diffusion-theory solution we can either use the diffusion equation to calculate the flux at a specific radius within each shell, such as the mid-point of the shell, or we can integrate the diffusion solution over the shell volume and calculate the volume-averaged flux. In the latter case, we obtain for the average flux between radii $[r_1, r_2]$

$$\phi_{avg}(r_1, r_2) = \frac{\int_{r_1}^{r_2} \phi(r) 4\pi r^2 dr}{\int_{r_1}^{r_2} 4\pi r^2 dr} = \frac{3\Sigma_t S}{4\pi}\left(\frac{3(r_2^2 - r_1^2)}{2(r_2^3 - r_1^3)} - \frac{1}{R}\right) \tag{5.1}$$

where $\phi(r)$ is given by eqn 3.34P. Here Σ_t is the total cross section of the scattering material, S is the source strength, and R is the extrapolated outer radius of the sphere. For a discussion of the extrapolated radius see p. 73 of the *Primer*. In order to compare our Monte Carlo result with the diffusion-theory solution, both eqn 3.34P and eqn 5.1 have been coded into the modified subroutine 'Stats,' with the values of r_1 and r_2 set equal to the radial boundaries of each of the scoring bins.

Table 5.2. Subroutine 'Col' for Exercise 5.1

```
SUBROUTINE COL                                                              !  1
REAL(8) FLTRN,delta                                                         !  2
REAL(8) x,y,z,u,v,w,xo,yo,zo,uo,vo,wo,wate,age,energ                        !  3
COMMON/PART/x,y,z,u,v,w,xo,yo,zo,uo,vo,wo,wate,age,energ,nzcur,newzn,ngroup !  4
REAL(8) dmfp,dtr,xsec,dcur                                                  !  5
COMMON/TRACK/dmfp,dtr,xsec,dcur                                             !  6
REAL(8) sigt(20),c(20);  ! dimensions allow up to 20 different media          7
COMMON/GEOM/sigt,c; !sigt is total cross section, c is non absorption prob    8
REAL(8) bscore(10),bsumsq(10),cscore(150),csumsq(150),bpart(10),cpart(150)
COMMON/STAT/bscore,bsumsq,cscore,csumsq,bpart,cpart
delta=dmfp/xsec                      ! distance traveled to collision         9
dtr=dcur+delta                       ! update total distance traveled        10
x=xo+u*dtr;  y=yo+v*dtr;  z=zo+w*dtr        ! update position                11
r=DSQRT(x*x+y*y+z*z)
i=INT(10.0d0*r)+1        ! index to zone in which collision occurs
cpart(i)=cpart(i)+wate  ! score wate of particle colliding in zone
wate=wate*c(nzcur)                   ! reduce wate by non-absorption probability 12
IF(wate.LT.0.1d0)THEN                ! if wate small play Russian roulette   13
   IF(wate.LT.FLTRN())THEN           ! if particle killed by RR              14
      nzcur=-1                       ! set nzcur=-1 to show particle killed  15
      RETURN                                                              ! 16
   ENDIF                                                                  ! 17
   wate=1.0d0                        ! particle survived RR, increase wate   18
ENDIF                                                                     ! 19
CALL ISOOUT                          ! assumes isotropic scatter in lab system 20
RETURN                                                                    ! 21
END                                                                       ! 22
```

Because we use a total cross section equal to one, the cross section data for this exercise, as contained in the file 'xsects.txt,' is identical to that of Example 5.1, shown in Table 5.24P. However, the geometry description for this exercise requires the outer radius of the sphere to be set to 15 cm. Thus the file 'geom.txt' shown in Table 5.23P must be modified such that the outer radius of SPH number one is 15 cm, as shown in Table 5.4.

Table 5.3. Subroutine 'Stats' for Exercise 5.1

```
SUBROUTINE Stats                                                              !     1
     REAL(8) bscore(10),bsumsq(10),cscore(150),csumsq(150),bpart(10),cpart(150)
     COMMON/STAT/bscore,bsumsq,cscore,csumsq,bpart,cpart                      !     3
     COMMON/IN/npart,nbatch                                                   !     4
     REAL(8) x,y,z,u,v,w,xo,yo,zo,uo,vo,wo,wate,age,energ                     !     5
     COMMON/PART/x,y,z,u,v,w,xo,yo,zo,uo,vo,wo,wate,age,energ,nzcur,newzn,ngroup  !  6
     REAL(8) tmp,tmp1,tmp2,var,stdev,pi,vol
ENTRY StatOne           ! entry point to initialize arrays for complete problem   8
     cscore=0.0d0;  csumsq=0.0d0
RETURN                                                                        !    10
ENTRY StatLp            ! entry point to initialize arrays for a particle         11
     cpart=0.0d0
RETURN                                                                        !    13
ENTRY StatELp           ! entry point to store scores for a particle              14
     DO i=1,150
       cscore(i)=cscore(i)+cpart(i)      !  store score
       csumsq(i)=csumsq(i)+cpart(i)**2   !  store square for variance calculation
     END DO
RETURN                                                                        !    19
ENTRY StatEnd           ! entry point to calculate and print results              20
     pi=2.0d0*DACOS(0.0d0)       ! to obtain a double precision Pi
     tmp=DFLOAT(npart)                                                        !    21
     tmp1=3.0/(4.0*pi)
     DO i=1,150
       var=csumsq(i)/tmp-(cscore(i)/tmp)**2      ! variance of cscore distribution
       stdev=DSQRT(var)                          ! standard deviation of cscore distr
       tmp2=DFLOAT(i)/10.0d0
       vol=4.0d0*pi*(tmp2**3-(tmp2-0.1d0)**3)/3.0d0   ! volumes to normalize scores
       dif1=tmp1*(1.0/(FLOAT(i)/10.0)-1.0/15.71)
       dif2=tmp1*(1.0/((FLOAT(i)-0.5)/10.0)-1.0/15.71)
       r2=FLOAT(i)/10.0;  r1=FLOAT(i-1)/10.0
       dif3=tmp1*((3.0*(r2**2-r1**2))/(2.0*(r2**3-r1**3))-1.0/15.71)
       WRITE(16,10)cscore(i)/tmp/vol,stdev/DSQRT(tmp)/vol,dif1,dif2,dif3
     END DO
10   FORMAT(1x,5F14.8)
RETURN                                                                        !    28
     END                                                                      !    29
```

Table 5.4. Geometry Description for Exercise 5.1

0	There are no RPPs
1	There is one SPH
0.0d0 0.0d0 0.0d0 15.0d0	Origin and radius of SPH
2	Number of zones
1	Zone one has one body
1	Body "1" is the body
1	Zone two has one body
-1	Body"-1" is the body

A partial list of the results obtained from running this problem using the modified routines shown above with 10^6 start particles is shown in Table 5.5. Three diffusion results – eqn 3.34P evaluated at the mid-points and the outer edges of the scoring intervals, and the average diffusion result for each scoring interval from eqn 5.1 – are shown for comparison. The three sets of diffusion results are similar except near the source point.

Table 5.5. Results of Exercise 5.1

Interval	Radius (mfp)	PFC result	Std dev	Diffusion right bdry	Diffusion mid point	Diffusion average
1	0-0.1	25.849	0.082	2.372	4.759	3.566
30	2.9-3.0	6.576-2	1.1-4	6.438-2	6.573-2	6.572-2
60	5.9-6.0	2.490-2	3.7-5	2.459-2	2.493-2	2.493-2
90	8.9-9.0	1.147-2	1.7-5	1.133-2	1.148-2	1.148-2
120	11.9-12.0	4.771-3	7.6-6	4.698-3	4.781-3	4.781-3
150	14.9-15.0	6.678-4	1.8-6	7.193-4	7.725-4	7.725-4

Discussion

The volume-averaged diffusion-theory results from eqn 5.1, which provide the best comparison with the Monte Carlo results, are lower than the Monte Carlo results at radii less than about two mfp, but are very close to the Monte Carlo results for radii larger than two mfp. There are also minor differences between the two results at the boundary, where the extrapolated radius of eqn 3.34P provides an approximation for the vacuum boundary whereas the Monte Carlo results model the correct effect of the boundary.

Exercise 5.2

Statement of the problem

2. Repeat Example 5.1 using a uniformly distributed source throughout the volume of the sphere, instead of a point source.

Solution

The selection of the radius for a point chosen randomly within a unit sphere is generally done by using eqn 2.18P. For a sphere of radius 10 cm, as is specified in Example 5.1, the radius given by eqn 2.18P must be multiplied by ten in order to distribute the start particles uniformly throughout the problem geometry. By symmetry, the flux in this problem depends only on the radius. Therefore once a start particle radius is selected the choice of the position of the start particle on a spherical shell of this radius is arbitrary. That is, as we have done in other problems with spherical symmetry, we are free to place all of the start particles along a single line, such as the +Z axis. On the other hand, the assumptions used in defining the source location in Exercise 3.2 could also be used in this exercise.

For the purpose of illustrating the general problem of distributing source particles uniformly within a spherical volume, and to follow strictly the statement of the exercise, we distribute the source uniformly throughout the sphere instead of along a single line. For this approach, once a source radius is selected, a random unit vector is used to determine the direction from the origin to the source point, as given by eqn 3.29P, and thus to define a random location for the source point in three-dimensional space. The direction cosines of the starting direction for each particle are chosen by applying eqn 3.29P a second time, as is done in subroutine 'Isoout,' shown in Table 5.12P.

The subroutines modified for Exercise 5.2 are listed in Table 5.6. The changes described above have been made in subroutine 'Source,' as shown in Table 5.7. That is, 'Isoout' is called twice, first to obtain a random direction from the origin to the source point, and second to obtain the direction cosines for the start particle.

Table 5.6. Subroutines used in Exercise 5.2

Subroutine	Location
'Source'	Table 5.7
'Col'	Table 5.8
'Stats'	Table 5.9

Table 5.7. Subroutine 'Source' for Exercise 5.2

```
SUBROUTINE SOURCE                                                            !   1
REAL(8) x,y,z,u,v,w,xo,yo,zo,uo,vo,wo,wate,age,energ                         !   2
REAL(8) fltrn,r
COMMON/PART/x,y,z,u,v,w,xo,yo,zo,uo,vo,wo,wate,age,energ,nzcur,newzn,ngroup  !   3
pi=2.0d0*DACOS(0.0d0)     ! keep for more complicated source that needs pi      4
r=10.d0*fltrn()**(1.0d0/3.0d0)
CALL ISOOUT
x=u*r;  y=v*r;  z=w*r
nzcur=1                   ! assumes source is in zone one                       6
CALL ISOOUT               ! direction chosen isotropically                      7
wate=1.0d0                ! particle starts with a weight of one                8
RETURN                                                                       !   9
END                                                                          !  10
```

As in Example 5.1, subroutine 'Col' is used to score the number of particle collisions that occur in each of 100 spatial bins. However, the version of this subroutine used in Example 5.1 has been modified to avoid possible problems caused by rounding errors associated with the initial source location. Given the coding in the modified subroutine 'Source,' it is possible that some particles will start slightly outside the 10-cm sphere. To ensure that such particles do not exceed array bounds, a test of the array subscript has been added to subroutine 'Col,' as shown in Table 5.8.

Although not requested in the exercise, it is useful to compare the Monte Carlo results with those obtained from the diffusion equation. In this case we have a distributed source and we cannot use eqn 3.34P. In spherical coordinates the steady-state diffusion equation with no absorption can be written

$$D\frac{d^2\phi}{dr^2}+\frac{2D}{r}\frac{d\phi}{dr}+S=0$$

where S is the source per unit volume and r is the radius. Because in this problem $\Sigma_t = \Sigma_s$, from the footnote on p. 106 of the *Primer* we have

$$D=\frac{1}{3\Sigma_s}$$

Table 5.8. Subroutine 'Col' for Exercise 5.2

```
SUBROUTINE COL                                                        !    1
REAL(8) FLTRN,delta                                                   !    2
REAL(8) x,y,z,u,v,w,xo,yo,zo,uo,vo,wo,wate,age,energ                  !    3
COMMON/PART/x,y,z,u,v,w,xo,yo,zo,uo,vo,wo,wate,age,energ,nzcur,newzn,ngroup  !  4
REAL(8) dmfp,dtr,xsec,dcur                                            !    5
COMMON/TRACK/dmfp,dtr,xsec,dcur                                       !    6
REAL(8) sigt(20),c(20);  !  dimensions allow up to 20 different media      7
COMMON/GEOM/sigt,c; !sigt is total cross section, c is non-absorption prob  8
REAL(8) bscore(10),bsumsq(10),cscore(100),csumsq(100),bpart(10),cpart(100)
COMMON/STAT/bscore,bsumsq,cscore,csumsq,bpart,cpart
delta=dmfp/xsec                   !  distance traveled to collision        9
dtr=dcur+delta                    !  update total distance traveled       10
x=xo+u*dtr;  y=yo+v*dtr;  z=zo+w*dtr       !  update position             11
r=DSQRT(x*x+y*y+z*z)
i=INT(10.0d0*r)+1        !  index to zone in which collision occurs
IF(i.GT.100)i=100
cpart(i)=cpart(i)+wate !  score wate of particle colliding in zone
wate=wate*c(nzcur)                !  reduce wate by non-absorption probability  12
IF(wate.LT.0.1d0)THEN             !  if wate small play Russian roulette  13
   IF(wate.LT.FLTRN())THEN        !  if particle killed by RR             14
      nzcur=-1                    !  set nzcur=-1 to show particle killed  15
      RETURN                                                          !   16
   ENDIF                                                              !   17
   wate=1.0d0                     !  particle survived RR, increase wate  18
ENDIF                                                                 !   19
CALL ISOOUT                       !  assumes isotropic scatter in lab system  20
RETURN                                                                !   21
END                                                                   !   22
```

The solution to this equation for a uniform source, which is given in a number of texts and can be confirmed by substitution, is

$$\phi(r)=\frac{S}{6D}\left(R'^2-r^2\right)=\frac{S\Sigma_s}{2}\left(R'^2-r^2\right)=\frac{3\Sigma_s}{8\pi R^3}\left(R'^2-r^2\right) \tag{5.2}$$

In eqn 5.2, R is the radius of the sphere and R' is the radius of the sphere plus the extrapolation distance of approximately 0.71 mfp; i.e., as was discussed in conjunction with Example 3.1, the vacuum boundary condition is approximated by having the diffusion theory solution go to zero at the extrapolated boundary instead of at the actual boundary. To obtain the last expression in eqn 5.2 the source S is normalized such that the total source over the volume of the sphere is one particle. Equation 5.2 is solved in subroutine 'Stats' in order to facilitate comparison with the Monte Carlo solution. The modified subroutine 'Stats' is given in Table 5.9.

Although the diffusion solution of eqn 5.2 could be integrated to give a volume-averaged result, in this case the use of the mid-point of the shells provides a good estimate of such an average. From eqn 5.2 it can be seen that, in this distributed-source problem, the diffusion-theory flux approaches a constant as the radius approaches zero. Therefore, for this case the mid-point solution should be accurate even for scoring bins close to the origin. Results obtained from running PFC with these modified routines and with 10^7 start particles are shown in Table 5.10.

Table 5.9. Subroutine 'Stats' with Diffusion Theory Results for Exercise 5.2

```
SUBROUTINE Stats                                                          !  1
    REAL(8) bscore(10),bsumsq(10),cscore(100),csumsq(100),bpart(10),cpart(100)
    COMMON/STAT/bscore,bsumsq,cscore,csumsq,bpart,cpart                   !  3
    COMMON/IN/npart,nbatch                                                !  4
    REAL(8) x,y,z,u,v,w,xo,yo,zo,uo,vo,wo,wate,age,energ                  !  5
    COMMON/PART/x,y,z,u,v,w,xo,yo,zo,uo,vo,wo,wate,age,energ,nzcur,newzn,ngroup  ! 6
    REAL(8) tmp,tmp1,tmp2,var,stdev,pi,vol,diff
ENTRY StatOne              ! entry point to initialize arrays for complete problem 8
    cscore=0.0d0; csumsq=0.0d0
RETURN                                                                    ! 10
ENTRY StatLp               ! entry point to initialize arrays for a particle    11
    cpart=0.0d0
RETURN                                                                    ! 13
ENTRY StatELp              ! entry point to store scores for a particle         14
    DO i=1,100
      cscore(i)=cscore(i)+cpart(i)      ! store score
      csumsq(i)=csumsq(i)+cpart(i)**2   ! store square for variance calculation
    END DO
RETURN                                                                    ! 19
ENTRY StatEnd              ! entry point to calculate and print results         20
    pi=2.0d0*DACOS(0.0d0)         ! to obtain a double precision Pi
    tmp=DFLOAT(npart)                                                     ! 21
    DO i=1,100
      var=csumsq(i)/tmp-(cscore(i)/tmp)**2        ! variance of cscore distribution
      stdev=DSQRT(var)                            ! standard deviation of cscore distr
      tmp2=DFLOAT(i)/10.0d0
      vol=4.0d0*pi*(tmp2**3-(tmp2-0.1d0)**3)/3.0d0  ! volumes to normalize scores
      diff=3.0d0/(8.0d0*pi*1000.0d0)*(10.71**2-(tmp2-0.05d0)**2)
      WRITE(16,10)cscore(i)/tmp/vol,stdev/DSQRT(tmp)/vol,diff
    END DO
10  FORMAT(1x,5F14.8)
RETURN                                                                    ! 28
    END                                                                   ! 29
```

Table 5.10. Results for Exercise 5.2

Interval	Radius (mfp)	PFC result	Std dev	Diffusion
1	0-0.1	0.01394	6.2-4	0.01369
2	0.1-0.2	0.01352	2.4-4	0.01369
3	0.2-0.3	0.01368	1.5-4	0.01368
4	0.3-0.4	0.01344	1.1-4	0.01368
5	0.4-0.5	0.01365	0.9-4	0.01367
20	1.9-2.0	0.01257	1.7-5	0.01265
40	3.9-4.0	0.01183	1.2-5	0.01183
60	5.9-6.0	0.009462	7.2-6	0.009466
80	7.9-8.0	0.006130	4.0-6	0.006248
100	9.9-10.0	0.001547	1.3-6	0.001874

Discussion

This exercise assumes a source distributed uniformly over the problem geometry. Therefore, most of the source particles are born at large radii. As a result, the number of collisions per start particle that occur within the volumes of the scoring regions near the origin is small. That is, because of the small size of the scoring regions near the origin, a significant number of start particles is required in order to obtain good results at these locations. Even with the 10^7 start particles used in this calculation there are relatively large statistical fluctuations in the first few scoring bins, as is evident from the results shown in Table 5.10. These fluctuations are a result of the small number of collisions that were tallied in these bins. This is an important conclusion that we will find repeated many times in problems in which most or all of the start particles are born far from the point at which we desire an answer, or when the size of our scoring volume is small compared with the size of the problem geometry. Biased sampling, which is discussed in Chapter 6 of the *Primer*, usually offers a solution to this problem.

The Monte Carlo and diffusion-theory results compare well in this exercise except near the outer boundary of the scattering material. We know that the extrapolated boundary condition used to approximate a vacuum boundary for the diffusion equation is only an approximate match to the correct boundary condition. We do not expect accurate diffusion results near such a boundary.

Exercise 5.3

Statement of the problem

3. Repeat Exercise 1 of Chapter 4 using PFC.

Solution

Exercise 4.1 involves the point source and spherical geometry of Examples 3.1 and 5.1, and assumes a pure scattering material. However, instead of assuming isotropic scattering in the laboratory system, it assumes isotropic scattering in the center-of-mass system while neglecting energy changes, with the scattering targets being assigned a mass of one atomic mass unit. The routine developed to solve Exercise 4.1 by calculating the scattering angles for isotropic scattering in the center-of-mass system, based on subroutine 'Isocoll' of Table 4.5P, is shown in Table 4.1. Changes in this routine and in subroutine 'Col' are made in order to solve this exercise, as listed in Table 5.11.

Table 5.11. Modified Subroutines for Exercise 5.3

Subroutine	Location
'Isocoll'	Table 5.12
'Col'	Table 5.13
'Stats'	Table 5.26P

For the present exercise a new version of the center-of-mass scattering routine, which is called 'Isoout' in Exercise 4.1, is used. This subroutine, shown in Table 5.12, is nearly identical to that given in Table 4.1. We have employed a new subroutine name, 'Isocoll,' since the name 'Isoout' is needed in other parts of PFC. In addition we use common statements to pass variables to the subroutine.

The particle scattering analysis used in Table 4.4P is modeled in the library version of subroutine 'Col.' We modify 'Col' to score the weights of the scattered particles in each spatial bin. In the revised version of this routine, shown in Table 5.13, the subroutine call in statement 20 of the library routine is changed to call the new subroutine 'Isocoll' rather than 'Isoout.' The geometry and cross section files used in this calculation are those of Example 5.1, shown in Tables 5.23P and 5.24P.

Selected results for Exercise 5.3, for a calculation using 10^4 start particles, are shown in Table 5.14. Shown for comparison are comparable results from Example 3.1 and Exercise 4.1. Although statistical uncertainties in these results are not shown, we could estimate them using the technique discussed under Exercise 4.1, pp. 39–41. The present calculation does not use the random number string in the same way as the Fortran for Exercise 4.1 and therefore we cannot reproduce the earlier results exactly. However, we see that the present results compare well with those for Exercise 4.1. We conclude that we have successfully reproduced that calculation using PFC. As expected, the results of Example 3.1, which assume isotropic scattering in the laboratory system, are uniformly higher than the center-of-mass scattering results.

Table 5.12. Subroutine 'Isocol1' for Exercise 5.3

```
      SUBROUTINE ISOCOL1
!     Subroutine ISOCOL1 of Table 4.5 of Primer.
!     Modified for Exercise 5.3. A is fixed as 1
!
      REAL(8) x,y,z,u,v,w,xo,yo,zo,uo,vo,wo,wate,age,energ
      COMMON/PART/x,y,z,u,v,w,xo,yo,zo,uo,vo,wo,wate,age,energ,nzcur,newzn,ngroup
      REAL(8) fltrn,PI,COSTH,PHI,UL,VL,WL,UF,VF,WF,SR,SINTH,T,A
      A=1.0d0
      PI=2.0d0*DACOS(0.0d0)
      COSTH = 2.D0*fltrn()-1.0D0  ! Cosine of random polar scatter angle selected
      PHI = fltrn()*2.D0*PI       ! Random azimuthal angle selected
      WL = (A*COSTH+1.D0)/DSQRT(A**2+2.D0*A*COSTH+1.D0) ! z-dir cosine,lab coordinates
      SINTH = DSQRT(1.D0-WL**2)   ! Azimuthal angle in lab coordinates
      UL = SINTH*DCOS(PHI)        ! direction cosine in lab coordinates
      VL = SINTH*DSIN(PHI)        ! direction cosine in lab coordinates
      IF(ABS(U).LT.0.9)THEN         ! select appropriate transformation
        SR=DSQRT(1.0D0-U*U)         ! Primary transformation T sub x
        UF=SR*UL+U*WL               ! final direction cosine U
        VF=-U*V*UL/SR+W*VL/SR+V*WL  ! final direction cosine V
        WF=-U*W*UL/SR-V*VL/SR+W*WL  ! final direction cosine W
     ELSE
        SR=DSQRT(1.0D0-V*V)         ! Alternate transformation T sub y
        UF=W*UL/SR-U*V*VL/SR+U*WL   ! final direction cosine U
        VF=SR*VL+V*WL               ! final direction cosine V
        WF=-U*UL/SR-V*W*VL/SR+W*WL  ! final direction cosine W
      ENDIF
      T=DSQRT(UF*UF+VF*VF+WF*WF)  ! renormalize to fix possible roundoff errors
      U=UF/T; V=VF/T; W=WF/T      ! set values for return
      RETURN ; END
```

Table 5.13. Subroutine 'Col' for Exercise 5.3

```
      SUBROUTINE COL                                                      !     1
      REAL(8) FLTRN,delta                                                 !     2
      REAL(8) x,y,z,u,v,w,xo,yo,zo,uo,vo,wo,wate,age,energ                !     3
      COMMON/PART/x,y,z,u,v,w,xo,yo,zo,uo,vo,wo,wate,age,energ,nzcur,newzn,ngroup  !  4
      REAL(8) dmfp,dtr,xsec,dcur                                          !     5
      COMMON/TRACK/dmfp,dtr,xsec,dcur                                     !     6
      REAL(8) sigt(20),c(20);  !  dimensions allow up to 20 different media     7
      COMMON/GEOM/sigt,c; !sigt is total cross section, c is non-absorption prob  8
      REAL(8) bscore(10),bsumsq(10),cscore(100),csumsq(100),bpart(10),cpart(100)
      COMMON/STAT/bscore,bsumsq,cscore,csumsq,bpart,cpart
      delta=dmfp/xsec                  !  distance traveled to collision        9
      dtr=dcur+delta                   !  update total distance traveled       10
      x=xo+u*dtr;  y=yo+v*dtr;  z=zo+w*dtr       !  update position            11
      r=DSQRT(x*x+y*y+z*z)
      i=INT(10.0d0*r)+1        !  index to zone in which collision occurs
      cpart(i)=cpart(i)+wate !  score wate of particle colliding in zone
      wate=wate*c(nzcur)               !  reduce wate by non-absorption probability  12
      IF(wate.LT.0.1d0)THEN            !  if wate small play Russian roulette  13
         IF(wate.LT.FLTRN())THEN       !  if particle killed by RR             14
             nzcur=-1                  !  set nzcur=-1 to show particle killed 15
             RETURN                                                       !    16
         ENDIF                                                            !    17
         wate=1.0d0                    !  particle survived RR, increase wate  18
      ENDIF                                                               !    19
      CALL ISOCOL1                     !  isotropic scatter in CM system with A=1
      RETURN                                                              !    21
      END                                                                 !    22
```

Table 5.14. Selected Results for Exercise 5.3

Radius (mfp)	Example 3.1	Exercise 4.1	Exercise 5.3
0.1	1114	1059	982
1	3055	1306	1293
2	4812	1783	1846
3	6386	2368	2348
4	7712	2716	2717
5	8205	3037	2972
6	7954	3157	3244
7	7265	2964	3023
8	6127	2838	2850
9	4473	2467	2491
10	1808	1679	1662

Exercise 5.4

Statement of the problem

4. Consider Example 3.5.
   a. Repeat the example using PFC. [Hint: Use an RPP with large x and y dimensions to approximate an infinite slab.]
   b. Convert the flight path determination in subroutine 'Dist' to the option of obtaining a new flight path whenever a particle track encounters a new zone. Repeat the example using this option in PFC.

Solution Part a

Example 3.5 involves a two-region slab with a spherical void, the center of the latter being located at the interface between the two regions. The void has a diameter equivalent to one mfp in the slab materials. As specified in Example 3.5, the compositions of the two layers are taken from Example 3.3. In that example the total cross sections of the two layers are the same, but the first layer is two mfp thick and is a pure scatterer while the second layer is one mfp thick and is a pure absorber. A monoenergetic, pencil beam of neutrons is normally incident upon the face of the scattering layer such that an extension of the beam passes through the center of the void, as shown in Figure 3.5P. The PFC subroutines modified for this calculation are shown in Table 5.15.

Table 5.15. Modified Subroutines for Exercise 5.4a

Subroutine	Location
'Bdrx'	Table 5.16
'Col'	Table 5.17
'Source'	Table 5.18
'Stats'	Table 5.19

The modified subroutine 'Bdrx' is shown in Table 5.16. This subroutine is used to score the reflected and transmitted neutrons. Two elements of the array 'bpart' are used in order to accommodate these scores. The modified subroutine 'Col' is shown in Table 5.17. This routine is used to score the number of incident particles that are absorbed. The scoring is done in the array 'cpart.' The modified subroutine 'Source,' shown in Table 5.18, provides a normally incident source of neutrons on the slab along the +Z axis. Finally, subroutine 'Stats,' shown in Table 5.19, has been modified to accumulate the desired reflection and transmission scores and to calculate the results.

Table 5.16. Subroutine 'Bdrx' for Exercise 5.4a

```
SUBROUTINE BDRX                                                                  !   1
REAL(8) delta                                                                    !   2
REAL(8) sphcx(20),sphcy(20),sphcz(20),sphr(20),rpxmin(20),rpxmax(20),&           !   3
 rpymin(20),rpymax(20),rpzmin(20),rpzmax(20),bdin(40),bdout(40)                  !   4
LOGICAL bdtest(40), bdhit(40)                                                    !   5
COMMON/cg/sphcx,sphcy,sphcz,sphr,rpxmin,rpxmax,rpymin,rpymax,rpzmin,rpzmax,&     !   6
 bdin,bdout,nsph,nrpp,nbdy,nzones,nbz(40,41),nnext(40,40,40,2),bdtest,bdhit      !   7
REAL(8) x,y,z,u,v,w,xo,yo,zo,uo,vo,wo,wate,age,energ                             !   8
COMMON/PART/x,y,z,u,v,w,xo,yo,zo,uo,vo,wo,wate,age,energ,nzcur,newzn,ngroup      !   9
REAL(8) dmfp,dtr,xsec,dcur                                                       !  10
COMMON/TRACK/dmfp,dtr,xsec,dcur                                                  !  11
REAL(8) bscore(10),bsumsq(10),cscore(10),csumsq(10),bpart(10),cpart(10)
COMMON/STAT/bscore,bsumsq,cscore,csumsq,bpart,cpart
delta=dtr-dcur                  ! delta--distance traveled to reach boundary         12
dcur=dtr                        ! update current distance traveled                   13
dmfp=dmfp-delta*xsec            ! subtract current distance in mfp from dmfp         14
x=x+u*delta; y=y+v*delta; z=z+w*delta     ! update position                          15
IF(nzcur.EQ.1.AND.z.LT.0.1)bpart(1)=wate  ! reflection
IF(nzcur.EQ.2.AND.z.GT.2.9)bpart(2)=wate  ! transmission
nzcur=newzn                     ! change identifier of current zone                  16
if(nzcur.LT.nzones)RETURN       ! if not at outer zone then return                   17
nzcur=-1                        ! set nzcur=-1 if in outer zone (for escape)         18
RETURN                                                                           !  19
END                                                                              !  20
```

The geometry description used for this calculation is shown in Table 5.20. The x and y dimensions of the RPPs used to define the geometry are chosen such that the respective edges of these bodies are 50 mfp from the Z axis. Since the incident source is along the Z axis, this is essentially equivalent to an infinite slab in these two dimensions. The question of the proper definition of lateral dimensions for modeling an infinite slab is discussed in Example 6.1. The cross section information for this exercise is shown in Table 5.21. The total cross sections for the slabs are both set to one, to correspond to the specified thicknesses of the slabs in mfp for the geometry of Table 5.20. The value of the non-absorption probability c is set to one for the pure scatterer and to zero for the pure absorber. The internal void is specified by a total cross section of zero.

Table 5.17. Subroutine 'Col' for Exercise 5.4a

```
      SUBROUTINE COL                                                           !  1
      REAL(8) FLTRN,delta,abs
      REAL(8) x,y,z,u,v,w,xo,yo,zo,uo,vo,wo,wate,age,energ                     !  3
      COMMON/PART/x,y,z,u,v,w,xo,yo,zo,uo,vo,wo,wate,age,energ,nzcur,newzn,ngroup   !  4
      REAL(8) dmfp,dtr,xsec,dcur                                               !  5
      COMMON/TRACK/dmfp,dtr,xsec,dcur                                          !  6
      REAL(8) sigt(20),c(20);  ! dimensions allow up to 20 different media        7
      COMMON/GEOM/sigt,c; !sigt is total cross section, c is non-absorption prob   8
      REAL(8) bscore(10),bsumsq(10),cscore(10),csumsq(10),bpart(10),cpart(10)  !
      COMMON/STAT/bscore,bsumsq,cscore,csumsq,bpart,cpart                      !
      delta=dmfp/xsec                 ! distance traveled to collision            9
      dtr=dcur+delta                  ! update total distance traveled           10
      x=xo+u*dtr;  y=yo+v*dtr;  z=zo+w*dtr     ! update position                 11
      abs=wate*(1.0d0-c(nzcur))       ! weight of particle absorbed
      cpart(1)=cpart(1)+abs           ! score absorbed weight
      wate=wate*c(nzcur)              ! reduce wate by non-absorption probability 12
      IF(wate.LT.0.1d0)THEN           ! if wate small play russian roulette      13
         IF(wate.LT.FLTRN())THEN      ! if particle killed by RR                  14
            nzcur=-1                  ! set nzcur=-1 to show particle killed      15
            RETURN                                                             ! 16
         ENDIF                                                                 ! 17
         wate=1.0d0                   ! particle survived RR, increase wate       18
      ENDIF                                                                    ! 19
      CALL ISOOUT                     ! assumes isotropic scatter in lab system   20
      RETURN                                                                   ! 21
      END                                                                      ! 22
```

Table 5.18. Subroutine 'Source' for Exercise 5.4a

```
      SUBROUTINE SOURCE                                                        !  1
      REAL(8) x,y,z,u,v,w,xo,yo,zo,uo,vo,wo,wate,age,energ                     !  2
      COMMON/PART/x,y,z,u,v,w,xo,yo,zo,uo,vo,wo,wate,age,energ,nzcur,newzn,ngroup   !  3
      pi=2.0d0*DACOS(0.0d0)        ! keep for more complicated source that needs pi  4
      x=0.0d0;  y=0.0d0;  z=0.1d-10   ! starts at origin just inside zone 1
      nzcur=1                      ! assumes origin is in zone one                 6
      u=0.0d0;  v=0.0d0;  w=1.0d0     ! starts into slab
      wate=1.0d0                   ! particle starts with a weight of one          8
      RETURN                                                                   !  9
      END                                                                      ! 10
```

Table 5.19. Subroutine 'Stats' for Exercise 5.4a

```
   SUBROUTINE Stats                                                            !  1
      REAL(8) bscore(10),bsumsq(10),cscore(10),csumsq(10),bpart(10),cpart(10)  !  2
      COMMON/STAT/bscore,bsumsq,cscore,csumsq,bpart,cpart                      !  3
      COMMON/IN/npart,nbatch                                                   !  4
      REAL(8) x,y,z,u,v,w,xo,yo,zo,uo,vo,wo,wate,age,energ                     !  5
      COMMON/PART/x,y,z,u,v,w,xo,yo,zo,uo,vo,wo,wate,age,energ,nzcur,newzn,ngroup   !  6
      REAL(8) tmp,tmp1,tmp2,var,stdev                                          !  7
   ENTRY StatOne          ! entry point to initialize arrays for complete problem  8
      bscore=0.0d0;  bsumsq=0.0d0;  cscore=0.0d0;  csumsq=0.0d0                !  9
   RETURN                                                                      ! 10
   ENTRY StatLp           ! entry point to initialize arrays for a particle       11
      bpart=0.0d0;  cpart=0.0d0                                                ! 12
   RETURN                                                                      ! 13
   ENTRY StatELp          ! entry point to store scores for a particle            14
      bscore(1)=bscore(1)+bpart(1)     ! store score                              15
      bsumsq(1)=bsumsq(1)+bpart(1)**2  ! store square for variance calculation    16
      bscore(2)=bscore(2)+bpart(2)     ! store score
      bsumsq(2)=bsumsq(2)+bpart(2)**2  ! store square for variance calculation
      cscore(1)=cscore(1)+cpart(1)     ! store score                              17
      csumsq(1)=csumsq(1)+cpart(1)**2  ! store square for variance calculation    18
   RETURN                                                                      ! 19
   ENTRY StatEnd          ! entry point to calculate and print results            20
```

Table 5.19. con't...

```
        tmp=DFLOAT(npart)                                     !                              21
        var=bsumsq(1)/tmp-(bscore(1)/tmp)**2                  !  variance of bscore distribution   22
        stdev=DSQRT(var)                                      !  standard deviation of bscore distr23
        WRITE(16,*)bscore(1)/tmp,stdev/DSQRT(tmp)             !  result and std dev of result      24
        var=bsumsq(2)/tmp-(bscore(2)/tmp)**2                  !  variance of bscore distribution
        stdev=DSQRT(var)                                      !  standard deviation of bscore distr
        WRITE(16,*)bscore(2)/tmp,stdev/DSQRT(tmp)             !  result and std dev of result
        var=csumsq(1)/tmp-(cscore(1)/tmp)**2                  !  variance of cscore distribution   25
        stdev=DSQRT(var)                                      !  standard deviation of cscore distr26
        WRITE(16,*)cscore(1)/tmp,stdev/DSQRT(tmp)             !  result and std dev of result      27
  RETURN                                                      !                              28
        END                                                   !                              29
```

Table 5.20. Geometry Description for Exercise 5.4a

2	There are two RPPs
-50.0 50.0 -50.0 50.0 0.0 2.0	Coordinates for RPP, body "1"
-50.0 50.0 -50.0 50.0 2.0 3.0	Coordinates for RPP, body "2"
1	There is one SPH
0.0 0.0 2.0 0.5	Origin and radius of SPH, body "3"
4	Number of zones
2	Zone one has two bodies
1 -3	Body "1" minus body "3"
2	Zone two has two bodies
2 -3	Body "2" minus body "3"
1	Zone three has one body
3	Body "3" is the body
2	Zone four has two bodies
-1 -2	Minus body "1" minus body "2"

Table 5.21. Description of Cross Sections for Exercise 5.4a

1.0 1.0	Non-absorption prob and total cross section - Zone 1
0.0 1.0	Non-absorption prob and total cross section - Zone 2
0.0 0.0	Non-absorption prob and total cross section - Zone 3

The results obtained from running this modified code using various numbers of incident particles are shown in Table 5.22. These results are essentially the same as those shown in Table 3.13P, with only statistical differences that arise because the random number sequences used are different. We conclude that the changes to PFC described above have correctly modeled the problem of Example 3.5.

Table 5.22. Results of Exercise 5.4a

No. source particles	Transmission	Reflection	Absorption
10^4	0.207 ± 0.004	0.485 ± 0.005	0.308 ± 0.005
10^5	0.2070 ± 0.0013	0.4926 ± 0.0016	0.3005 ± 0.0014
10^6	0.2070 ± 0.0004	0.4888 ± 0.0005	0.3042 ± 0.0005

Solution Part b

Exercise 5.4b calls for converting the flight path determination in subroutine 'Dist' to the option of obtaining a new flight path whenever a particle track encounters a new zone, then repeating Example 3.5 using this option in PFC. Subroutine 'Dist' is used to determine the flight path in terms of the number of mfp a neutron will travel. However, subroutine 'Walk' determines when a new path length is chosen. In the standard version of 'Walk' a new flight path is chosen for each start particle and after each collision. The collision results in a new direction of travel while the call to 'Dist' produces a new distance of travel in mfp. For this exercise the desired modification is to pick a new flight path length, in mfp, after each boundary crossing, in addition to doing so after each collision; hence the change is made in 'Walk' rather than 'Dist.'

A list of the subroutines modified for this calculation is shown in Table 5.23. The only change needed compared with the subroutines used in Exercise 5.4a is that in subroutine 'Walk,' shown in Table 5.24. In order to obtain a new flight path after each boundary crossing we have placed the call to 'Dist' inside the track loop.

Table 5.23. Modified Subroutines for Exercise 5.4b

Subroutine	Location
'Bdrx'	Table 5.16
'Col'	Table 5.17
'Source'	Table 5.18
'Stats'	Table 5.19
'Walk'	Table 5.24

The results obtained from running this modified code for various numbers of start particles are shown in Table 5.25. Comparing these results with those of Table 5.22 we see that, within statistics, selecting new flight paths for particles each time they cross a boundary in the geometry instead of continuing the original flight path does not change the result. This is consistent with the results obtained in Example 3.4.

Table 5.24. Subroutine 'Walk' for Exercise 5.4b

```
      SUBROUTINE WALK                                                    !      1
      COMMON/IN/npart,nbatch                                             !      2
      REAL(8) dmfp,dtr,xsec,dcur                                         !      3
      COMMON/TRACK/dmfp,dtr,xsec,dcur                                    !      4
      REAL(8) x,y,z,u,v,w,xo,yo,zo,uo,vo,wo,wate,age,energ               !      5
      COMMON/PART/x,y,z,u,v,w,xo,yo,zo,uo,vo,wo,wate,age,energ,nzcur,newzn,ngroup   !  6
Loop_Over_Particles: DO i=1,npart    ! loop over number of particles              7
      CALL StatLp                    ! initialize statistics for each particle    8
      CALL SOURCE                    ! get source parameters for particle         9
  Loop_for_Collisions: DO            ! to find collision sites                   10
    Loop_Track:  DO                  ! track particle to collision point         12
      CALL DIST                      ! get distance of travel in mfp
      CALL HIT                       ! get distance to boundary of current zone  13
      CALL MXSEC                     ! get total cross section for current zone  14
      IF((dtr-dcur)*xsec.GE.dmfp) EXIT Loop_Track  !  collision before boundary   15
      CALL BDRX                      ! process boundary crossing                 16
      IF(nzcur.EQ.-1) EXIT Loop_for_Collisions  !  particle killed e.g., by escaping  17
    END DO Loop_Track                                                    !       18
      CALL COL                       ! call col to process collision             19
      IF(nzcur.EQ.-1) EXIT Loop_for_Collisions  !  particle killed by collision   20
  END DO Loop_for_Collisions                                             !       21
      CALL StatELp                   ! process statistics when particle killed   22
END DO Loop_Over_Particles                                               !       23
      RETURN                                                             !       24
      END                                                                !       25
```

Table 5.25. Results for Exercise 5.4b

No. source particles	Transmission	Reflection	Absorption
10^4	0.206 ± 0.004	0.489 ± 0.005	0.305 ± 0.005
10^5	0.2063 ± 0.0013	0.4898 ± 0.0016	0.3040 ± 0.0015
10^6	0.2069 ± 0.0004	0.4888 ± 0.0005	0.3043 ± 0.0005

Chapter 6

Variance Reduction Techniques

Exercise 6.1

Statement of the problem

1. Use the angular-dependent results in Example 6.1 to devise a scheme for biasing the source angular distribution.

   a. Implement your source angular biasing and repeat the calculation while assuming a uniform spatial source distribution. What effect did this biasing have on the efficiency of your calculation?

   b. Combine your angular biasing with the spatial biasing used in the example and repeat the calculation employing both types of biasing simultaneously. What effect did the addition of angular biasing have on the efficiency of the calculation?

Solution

We defined a simple, spatial biasing technique, combined with stratification, as part of Example 6.1. The spatial biasing is prepared on the basis of the results of a stratified calculation that used no biasing. To estimate the bias parameters, the right leakage from each of the source strata obtained from the unbiased calculation is summed over the angle bins. These angle-integrated right leakage results are shown in Table 6.11P. To develop a scheme for angular biasing we use an analogous approach. That is, we re-run the unbiased, stratified calculation and sum the leakage results over the spatial bins, then use these results to set angular bias parameters. Because the analog results (the unbiased results without stratification) can be obtained with few modification to the Fortran we develop, the subroutines are modified to produce analog results in addition to the stratified results. The list of modified PFC subroutines required to perform this unbiased, stratified calculation is shown in Table 6.1. The geometry and cross section input data are identical to those of Example 6.1. The geometry file is shown in Table 6.6P. The cross section file uses the value c = 0.5 and $\Sigma_t = 10$ as described on p. 144 of the *Primer*.

Table 6.1. Modified Subroutines for Exercise 6.1 without Biasing

Subroutine	Location
'Col'	Table 6.4P
'Source'	Table 6.2
'Bdrx'	Table 6.3
'Stats'	Table 6.4

The only change made to the version of subroutine 'Source' shown in Table 6.8P is that necessary to change the dimensions of the arrays to accommodate the statistics. The modified subroutine is shown in Table 6.2. Subroutine 'Bdrx,' shown in Table 6.3, has been changed from that of Table 6.9P to score the analog results in addition to the stratified results.

Table 6.2. Subroutine 'Source' for Exercise 6.1 without Biasing

```
SUBROUTINE SOURCE                                                            !   1
REAL(8) x,y,z,u,v,w,xo,yo,zo,uo,vo,wo,wate,age,energ                         !   2
COMMON/PART/x,y,z,u,v,w,xo,yo,zo,uo,vo,wo,wate,age,energ,nzcur,newzn,ngroup  !   3
REAL(8) FLTRN
REAL(8) bscore(11,10),bsumsq(11,10),cscore(11,10),csumsq(11,10),&
 bpart(10),cpart(10)
COMMON/STAT/bscore,bsumsq,cscore,csumsq,bpart,cpart,&
 nsamp(10,10),nsampb(10,10),nsampc(10,10)
COMMON/EXAMPLE/n,m
nzcur=1                      ! source is in zone one
z=FLTRN()                    ! pick uniformly across slab
x=0.0d0;  y=0.0d0            ! starts particle along z-axis
CALL ISOOUT                  ! direction chosen isotropically                   7
wate=1.0d0                   ! particle starts with a weight of one             8
m=INT(5.0d0*(w+1.0d0))+1     ! index for scoring by angle
IF(m.GT.10)m=10
n=INT(10.0d0*z)+1            ! index for scoring by location
IF(n.GT.10)n=10
nsamp(n,m)=nsamp(n,m)+1      ! number of particles in each bin
RETURN                                                                       !   9
END                                                                          !  10
```

In order to determine the leakage as a function of source angle, four arrays have been added to subroutine 'Stats.' The modified subroutine is shown in Table 6.4. These arrays, 'astratb' and 'astratc,' along with their squares for estimating variances, 'astratb2' and astratc2,' are used to sum each of the angular-dependent leakage results over the ten spatial bins. The arrays are written to the output file at the end of the calculation along with analogous sums over the source angles. The latter results, from arrays 'zstratb' and zstratc,' are identical to those used in the unbiased, stratified calculation performed in Example 6.1, as shown in Table 6.10P.

Table 6.3. Subroutine 'Bdrx' for Exercise 6.1 without Biasing

```
SUBROUTINE BDRX                                                        !   1
REAL(8) delta,score
REAL(8) x,y,z,u,v,w,xo,yo,zo,uo,vo,wo,wate,age,energ                   !   8
COMMON/PART/x,y,z,u,v,w,xo,yo,zo,uo,vo,wo,wate,age,energ,nzcur,newzn,ngroup  !  9
REAL(8) dmfp,dtr,xsec,dcur                                             !  10
COMMON/TRACK/dmfp,dtr,xsec,dcur                                        !  11
REAL(8) bscore(11,10),bsumsq(11,10),cscore(11,10),csumsq(11,10),&
 bpart(10),cpart(10)
COMMON/STAT/bscore,bsumsq,cscore,csumsq,bpart,cpart,&
 nsamp(10,10),nsampb(10,10),nsampc(10,10)
COMMON/EXAMPLE/n,m
delta=dtr-dcur                   ! delta--distance traveled to reach boundary  12
dcur=dtr                         ! update current distance traveled            13
dmfp=dmfp-delta*xsec             ! subtract current distance in mfp from dmfp  14
z=z+w*delta                      ! update position in z-direction
score=0.01d0                     ! since bin should contribute 1/100 of source
IF(z.LE.0.0d0)THEN
 bpart(1)=wate
 nsampb(n,m)=nsampb(n,m)+1            ! leaks left
 bscore(n,m)=bscore(n,m)+score        ! store score
 bsumsq(n,m)=bsumsq(n,m)+score**2     ! store square for variance calculation
ELSE
 cpart(1)=wate
 nsampc(n,m)=nsampc(n,m)+1            ! leaks right
 cscore(n,m)=cscore(n,m)+score        ! store score
 csumsq(n,m)=csumsq(n,m)+score**2     ! store square for variance calculation
ENDIF
nzcur=-1                         ! set nzcur=-1 if in outer zone (for escape)  18
RETURN                                                                 !  19
END                                                                    !  20
```

We retain the ten angular bins used in Example 6.1. These bins are spaced evenly over the range of the cosine of the polar angle θ, measured with respect to the Z axis. That is, angle bin 1 corresponds to the interval $-1.0 < \cos\theta \le -0.8$, angle bin 2 to the interval $-0.8 < \cos\theta \le -0.6$, etc. Thus there are five bins in the negative-Z direction and five in the positive-Z direction. Intuitively one would expect the source particles born in the most forward-peaked (+Z) directions to be most important for scoring the right, or positive-Z, leakage. Similarly those born in the most backward-peaked directions should be least important for scoring the right leakage. Particles born in a direction nearly tangent to the surfaces of the slab should be less important to leakage through the nearest surface than those born at the same point in a direction aiming at the nearest surface.

The right leakage as a function of source angle, summed over the space points (the array 'astratc' in subroutine 'Stats'), as calculated by the unbiased, stratified model using 10^7 start particles, is given in Table 6.5. As expected, the results increase as the cosine of the start angle increases from -1 to $+1$. If we normalize the right leakage variance results of this table by dividing by the variance for angle bin 1, we obtain the results shown in the last column of Table 6.5. We use the results in this column to define our angular bias scheme by assigning approximate relative importances to the ten angle bins. Thus the relative importance of bins 1-4, for scoring the right leakage, is approximately one, while the values for bins 5-10 are roughly 2, 2, 3, 4, 5, and 5, respectively.

Table 6.4. Subroutine 'Stats' for Exercise 6.1 without Biasing

```
SUBROUTINE Stats                                                          !    1
   REAL(8) tmp,tmp1,tmp2,var,stdev,totalsb,totalsc,tvarb,tvarc,xnorm
   REAL(8) bscore(11,10),bsumsq(11,10),cscore(11,10),csumsq(11,10),&
     bpart(10),cpart(10),zstratb(10),zstratb2(10),zstratc(10),zstratc2(10),&
     astratb(10),astratb2(10),astratc(10),astratc2(10)
   COMMON/STAT/bscore,bsumsq,cscore,csumsq,bpart,cpart,&
     nsamp(10,10),nsampb(10,10),nsampc(10,10)
  COMMON/example/n,m
  COMMON/in/npart,nbatch
ENTRY StatOne              ! entry point to initialize arrays for complete problem  8
    bscore=0.0d0; bsumsq=0.0d0; cscore=0.0d0; csumsq=0.0d0              !    9
    nsamp = 0.; zstratb = 0.; zstratb2 = 0.; zstratc = 0.; zstratc2 = 0.
    astratb = 0.; astratb2 = 0.; astratc = 0.; astratc2 = 0.
RETURN                                                                  !   10
ENTRY StatLp               ! entry point to initialize arrays for a particle    11
     bpart=0.0d0; cpart=0.0d0                                           !   12
RETURN                                                                  !   13
ENTRY StatELp              ! entry point to store scores for a particle         14
     bscore(11,1)=bscore(11,1)+bpart(1)      ! store score
     bsumsq(11,1)=bsumsq(11,1)+bpart(1)**2   ! store square for var calculation
     cscore(11,1)=cscore(11,1)+cpart(1)      ! store score
     csumsq(11,1)=csumsq(11,1)+cpart(1)**2   ! store square for var calculation
RETURN                                                                  !   19
ENTRY StatEnd              ! entry point to calculate and print results         20
     WRITE(16,*)'analog results plus std dev for left then right leakage '
     tmp=DFLOAT(npart)                                                  !   21
     var=bsumsq(11,1)/tmp-(bscore(11,1)/tmp)**2        ! variance of bscore distr
     stdev=DSQRT(var)                               ! standard deviation of bscore distr23
     WRITE(16,*)bscore(11,1)/tmp,stdev/DSQRT(tmp)   ! result and std dev of result
     var=csumsq(11,1)/tmp-(cscore(11,1)/tmp)**2        ! variance of cscore distr
     stdev=DSQRT(var)                               ! standard deviation of cscore distr26
     WRITE(16,*)cscore(11,1)/tmp,stdev/DSQRT(tmp)   ! result and std dev of result
    totalsb = 0.; totalsc = 0.; tvarb = 0.; tvarc = 0.
    DO 200 i = 1,10
    DO 200 j = 1,10
     IF(nsamp(i,j).GE.1)THEN
       xnorm = DFLOAT(nsamp(i,j))
! left leakage totals and statistics
       bscore(i,j) = bscore(i,j)/xnorm             !left score by n,m
       var = (bsumsq(i,j)/xnorm-bscore(i,j)**2)/xnorm     !var of left score by n,m
       totalsb = totalsb + bscore(i,j)             !total left score
       zstratb(i) = zstratb(i)+bscore(i,j)         !left score in stratum i
       tvarb = tvarb + var                         !var of total left score
       zstratb2(i) = zstratb2(i)+ var              !var of stratum i left score
       bsumsq(i,j) = var                           !save var of left score by n,m
       astratb(j) = astratb(j)+bscore(i,j)         !left score in angle j
       astratb2(j) = astratb2(j)+ var              !var of angle j left score
! right leakage totals and statistics
       cscore(i,j) = cscore(i,j)/xnorm             !right score by n,m
       var = (csumsq(i,j)/xnorm-cscore(i,j)**2)/xnorm     !var of right score by n,m
       totalsc = totalsc + cscore(i,j)             !total right score
       zstratc(i) = zstratc(i)+cscore(i,j)         !right score in stratum i
       tvarc = tvarc + var                         !var of total right score
       zstratc2(i) = zstratc2(i) + var             !var of stratum i right score
       csumsq(I,j) = var                           !save var of right score by n,m
```

Table 6.4. con't...

```
        astratc(j) = astratc(j)+cscore(i,j)        !right score in angle j
        astratc2(j) = astratc2(j)+ var             !var of angle j right score
      ENDIF
  200 CONTINUE
      WRITE(16,111);  111 FORMAT(/,' left total leakage, std dev ')
      WRITE(16,106)totalsb,DSQRT(tvarb)
      WRITE(16,101);  101 FORMAT(/,' right total leakage, std dev ')
      WRITE(16,106)totalsc,DSQRT(tvarc);  106  FORMAT(1p2e14.4)
      WRITE(16,102);  102 FORMAT(/,' right leakage by bin',/,&
       ' angle bin        start wate         score            var   No. of scores')
      DO 300 i = 1,10
        WRITE(16,105)i;  105   FORMAT(/,' spatial bin i = ',i3)
      DO 300 j = 1,10
        WRITE(16,107)j, nsamp(i,j),cscore(i,j),csumsq(i,j),nsampc(i,j)
  107   FORMAT(i3,10x,i14,1p2e14.3,i14)
  300 CONTINUE
       WRITE(16,103)
    103 FORMAT(/,' left leakage by spatial bin summed over angle',/,&
       ' spatial bin      score           var           stdev')
       WRITE(16,108)(i,zstratb(i),zstratb2(i),DSQRT(zstratb2(i)),i=1,10)
    108 FORMAT(i3,10x,1p3e12.3)
       WRITE(16,109); 109 FORMAT(/,' right leakage by spatial bin summed over angle')
       WRITE(16,108)(i,zstratc(i),zstratc2(i),DSQRT(zstratc2(i)),i=1,10)
       WRITE(16,104); 104 FORMAT(/,' left leakage by angle summed over space points ')
       WRITE(16,108)(i,astratb(i),astratb2(i),DSQRT(astratb2(i)),i=1,10)
       WRITE(16,110); 110 FORMAT(/,' right leakage by angle summed over space points ')
       WRITE(16,108)(i,astratc(i),astratc2(i),DSQRT(astratc2(i)),i=1,10)
  RETURN                                                                  !     28
       END                                                                !     29
```

In order to examine the effects of angular biasing, we attempt to improve only the right leakage. Our candidate angle bias parameters are shown in Table 6.6. The second column of Table 6.6 shows the importance values we have assigned to each of the starting angular bins. By summing and normalizing these importances we obtain the values for the biased probability of a source particle being born in each angle bin shown in the third column of Table 6.6. We know the true probability of a source particle being born in each bin is 0.1, which is just the reciprocal of the number of angle bins we are using, because we have an isotropic source and we defined the angle bins to be equal in solid angle. The resulting ratio of the true probability to the biased probability of a source particle being born in each of the source angle bins is shown in the last column of Table 6.6.

The present selection of angular bias values is neither optimum nor unique. In particular, looking toward combining spatial and angular biasing, one could use the doubly differential results for the leakage (i.e., the leakage as a function of both space interval and angle interval) to derive a more sophisticated, and probably more efficient, set of bias parameters than that assumed here. Such a set of parameters could incorporate a different biased angular distribution for each spatial bin instead of a single biased angular distribution that is to be used for all of the spatial bins.

Table 6.5. Right Leakage for Exercise 6.1 without Biasing

Angular Stratum	Upper bound cosθ	Leak Right Probability	Leak Right Variance, Var_n	Ratio Var_n to Var_1
1	-0.8	1.038-3	9.874-11	1.00
2	-0.6	1.198-3	1.129-10	1.14
3	-0.4	1.330-3	1.241-10	1.26
4	-0.2	1.549-3	1.428-10	1.45
5	0.	1.895-3	1.707-10	1.73
6	0.2	3.106-3	2.539-10	2.57
7	0.4	5.144-3	3.444-10	3.49
8	0.6	7.133-3	4.150-10	4.20
9	0.8	9.104-3	4.940-10	5.00
10	1.0	1.113-2	5.773-10	5.85

Table 6.6. Angular Bias Parameters for Exercise 6.1

Angular Bin	Unnormalized Importance	Biased Probability	True Probability / Biased Probability
1	1	0.04	2.5
2	1	0.04	2.5
3	1	0.04	2.5
4	1	0.04	2.5
5	2	0.08	1.25
6	2	0.08	1.25
7	3	0.12	0.83333
8	4	0.16	0.625
9	5	0.20	0.5
10	5	0.20	0.5

It is possible to accommodate the present, spatially averaged, angular biasing in a manner analogous to that used for the spatial biasing in Example 6.1. However, because the last part of the present exercise calls for combining spatial and angular biasing it seems appropriate to consider a means for combining all of the options needed for executing Example 6.1, as well as for solving part a of Exercise 6.1, into one set of subroutines, and allowing the user to set the type of calculation desired by means of an input parameter. The subroutines that have been modified to accomplish this are listed in Table 6.7.

The modified subroutine 'Input' is shown in Table 6.8. A new input parameter 'ifbias' is read in this subroutine and is passed to the other subroutines through the 'example' common block. This common block is introduced in the spatially biased calculation of Example 6.1 and has been extended here to include the new input parameter. By specifying the 'ifbias' parameter as defined in subroutine 'Input,' the user can perform any of four sets of calculations. For each set, an analog calculation is performed as well as a stratified calculation. The four sets are: no biasing, spatial biasing, angular biasing, and a combination of spatial and angular biasing.

Table 6.7. Modified Subroutines for Exercise 6.1 with Biasing

Subroutine	Location
'Col'	Table 6.4P
'Bdrx'	Table 6.3
'Stats'	Table 6.4
'Input'	Table 6.8
'Source'	Table 6.9

Table 6.8. Subroutine 'Input' for Exercise 6.1 with Biasing

```
      SUBROUTINE INPUT                                                    !     1
      COMMON/IN/npart,nbatch                                              !     2
      COMMON/EXAMPLE/n,m,ifbias
      WRITE(*,12)                                                         !     3
   12 FORMAT(' Enter the number of particles.')                           !     4
      READ(*,*)npart                                                      !     5
      WRITE(16,*)'Number of particles = ',npart                           !     6
      WRITE(*,13)                                                         !     7
   13 FORMAT(' Enter the random number seed, an integer. ')               !     8
      READ(*,*)iseed                                                      !     9
      IF(iseed.LT.1)iseed=1                                               !    10
      IF(iseed.GE.2147483647)iseed=2147483646                             !    11
      WRITE(16,*)' Starting random number seed is ',iseed                 !    12
      CALL rndin(iseed)                                                   !    13
      WRITE(*,'(1x,a\)')&
       ' Type of bias (1,2,3,4)=(strat,spatial,angle,both)  '
      READ(*,*)ifbias
      WRITE(16,1)ifbias
    1 FORMAT(/,' Type of bias (1,2,3,4=strat,spatial,angle,both) = ',i3,/)
      nbatch=1                    ! needs to be changed if batches are to be used   14
      RETURN                                                              !    15
      END                                                                 !    16
```

The modified subroutine 'Source' is shown in Table 6.9. The options programmed into this version of the 'Source' routine are differentiated according to whether or not spatial biasing is to be used, and whether or not angular biasing is to be used. Particle weights are calculated in order to perform an unstratified calculation with the two biasing options. If only stratification is to be used then it is not necessary to calculate start weights. When angular biasing is used, this routine modifies the start weight of the particle depending on the angular bin in which the particle is born.

Once an angular bin has been selected in 'Source,' the particle is started in a random direction within that bin. This is accomplished by modifying the equation for the uniform distribution of source particles over a sphere into a uniform distribution of source particles over a spherical sector; i.e., over the portion of the 4π solid angle around the source point lying between the two polar angle limits of the bin. Because the area of the portion of the surface of a sphere between polar angles θ_1 and θ_2, where $\theta_1 < \theta_2$, is proportional to $\cos\theta_1 - \cos\theta_2$, the selection of a random polar angle θ between the limits (θ_1, θ_2) is obtained by selecting a polar cosine uniformly between $\cos\theta_1$ and $\cos\theta_2$ and taking the inverse cosine of the result. By symmetry the leakage will be independent of the azimuthal angle at which the particle is started and, for present purposes, the initial direction of the start particles will be in the X-Z plane.

Table 6.9. Subroutine 'Source' for Exercise 6.1

```
      SUBROUTINE SOURCE                                                        !   1
      REAL(8) x,y,z,u,v,w,xo,yo,zo,uo,vo,wo,wate,age,energ                      !   2
      COMMON/PART/x,y,z,u,v,w,xo,yo,zo,uo,vo,wo,wate,age,energ,nzcur,newzn,ngroup  !  3
      REAL(8) FLTRN
      REAL(8) bscore(11,10),bsumsq(11,10),cscore(11,10),csumsq(11,10),&
       bpart(10),cpart(10)
      COMMON/STAT/bscore,bsumsq,cscore,csumsq,bpart,cpart,&
       nsamp(10,10),nsampb(10,10),nsampc(10,10)
      COMMON/EXAMPLE/n,m,ifbias
      DIMENSION ibias(3,22)
      DATA ibias/0,0,4,0,0,4,0,0,4,0,0,4,&
       1,4,3,1,4,3,1,4,3,2,7,2,2,7,2,3,9,1,4,10,1,5,11,1,6,12,1,&
       7,13,2,7,13,2,8,15,3,8,15,3,8,15,3,&
       9,18,4,9,18,4,9,18,4,9,18,4/
      DIMENSION abias(10)
      DATA abias/0.04, 0.04, 0.04, 0.04, 0.08, 0.08, 0.12, 0.16, 0.20, 0.20/
      nzcur=1                          ! source is in zone one
      x=0.0d0;  y=0.0d0                ! starts particle along z-axis
      wate=1.0d0                       ! particle starts with a weight of one       8
    Check_SpatialBiasing: SELECT CASE(ifbias)
      CASE(1,3)                        ! no spatial biasing
        z=FLTRN()                      ! pick uniformly across slab
      CASE(2,4)                        ! spatial biasing
        r=FLTRN()*22.0d0
        i=INT(r)+1; IF(i.GT.22)i=22
        z=((r-DFLOAT(ibias(2,i)))/DFLOAT(ibias(3,i))+DFLOAT(ibias(1,i)))/10.0d0
        wate=2.2d0/DFLOAT(ibias(3,i))
    END SELECT Check_SpatialBiasing
    Check_AngularBiasing: SELECT CASE(ifbias)
      CASE(1,2)                        ! no angular biasing
        CALL ISOOUT                    ! direction chosen isotropically
      CASE(3,4)                        ! angular biasing
        r = fltrn()
        sum = 0.
        jangle = 10
        DO i = 1,10
          sum = abias(i)+sum
          IF (r.lt.sum) THEN
          jangle=i; EXIT
          ENDIF
        END DO
        alower = FLOAT(jangle-1)/5.-1.   ! find random cosine within biased interval
        w = alower + 0.2*fltrn()
        u = SQRT(1.-w*w)               ! start particle in x-z plane
        v = 0.
        wate=wate/(10.0d0*abias(jangle))
    END SELECT Check_AngularBiasing
      m=INT(5.0d0*(w+1.0d0))+1         ! index for scoring by angle
      IF(m.GT.10)m=10
      n=INT(10.0d0*z)+1                ! index for scoring by location
      IF(n.GT.10)n=10
      nsamp(n,m)=nsamp(n,m)+1          ! number of particles in each bin
      RETURN                                                                   !   9
      END                                                                      !  10
```

The left and right total leakage results obtained by running PFC with the preceding subroutines and the same geometry and cross section files used in the unbiased calculation, while exercising the various options discussed above for solving the problem, are shown in Table 6.10. All results are based on 10^8 start particles and all used a start random number of one. All of the calculations listed in Table 6.10 produced right leakage results that are in agreement with the results of Example 6.1.

Table 6.10. Results for Exercise 6.1

'ifbias'	Type of Calculation	Left Leakage	Standard Deviation	Right Leakage	Standard Deviation	Run Time (sec)	Efficiency to Right
1	analog, unstratified	4.266-2	2.021-5	4.268-2	2.021-5	599	4.09+6
1	stratified, unbiased	4.266-2	1.654-5	4.267-2	1.654-5	599	6.10+6
2	unstratified, spatial bias	4.264-2	1.663-5	4.267-2	1.664-5	606	5.96+6
2	stratified, spatial bias	4.265-2	1.438-5	4.266-2	1.438-5	606	7.98+6
3	unstratified, angular bias	4.266-2	2.984-5	4.267-2	1.899-5	572	4.85+6
3	stratified, angular bias	4.267-2	2.328-5	4.266-2	1.676-5	572	6.22+6
4	unstratified, both biases	4.267-2	2.472-5	4.267-2	1.559-5	586	7.02+6
4	stratified, both biases	4.267-2	2.044-5	4.266-2	1.431-5	586	8.33+6

As is discussed on p. 36 of the *Primer*, the accuracy to which one can estimate the uncertainty of a random variable is generally less than the accuracy to which the variable itself can be determined. Therefore large uncertainties in an estimated result, in terms of the fractional standard deviation (fsd), indicate that the variance is probably poorly known and the calculated result is suspect over and above the error implied by the variance estimate. As a rule of thumb it is recommended in the *Primer* that, for non-zero random variables with finite variances, a result with an fsd greater than about 0.1 should be considered questionable.

In the present case, where we are calculating efficiencies and are attempting to use several significant digits in the variance results in order to compare the efficiencies of different calculations, we need to obtain fsd values much less than 0.1. Good results could have been obtained with fewer particles than have been used both here and in subsequent calculations. The run times used are relatively long in order to obtain reliable estimates of the variance, and thus of the efficiency. The efficiencies are calculated using eqn 6.2P.

Although the efficiencies are shown in Table 6.10 to three significant digits, in general they are not known to this level of accuracy. Because we have specified the run times in integral seconds, there is an uncertainty of ±0.5 sec in the run times. This uncertainty produces a comparable uncertainty in the calculated efficiency. In addition, computer speeds and compilers vary, and it is possible that the coding style of the user could affect the run time. With a multitasking operating system the run time will vary with the number of programs in use. Therefore the location of the maximum in the efficiency could change if the same problem were executed on a different computer, if a different Fortran compiler were used, or if the number of open programs were changed. As a result, a user may find slightly different parameter values than those given here are required to obtain maximum efficiency.

We see from Table 6.10 that adding stratification to the analog case gives an increase in efficiency of 49%. Adding spatial biasing to the unstratified analog model increases the efficiency by 46%. As in Example 6.1, these improvements apply to both the left and right leakage results. If the spatial biasing were to be developed specifically for the right leakage, as has been done for the angular biasing model, the increase in efficiency should be larger for the right leakage result than that shown here. When the spatial and angular biasing techniques are used together in the unstratified model, the efficiency of the right leakage result is increased by 95%.

Adding angular biasing to the unstratified, analog case increases the efficiency of the right leakage by 19%, while the combination of angular biasing and stratification increases the efficiency of the right leakage result by 52%. Thus the calculation of the right leakage is only slightly improved by the use of angular biasing in combination with stratification, compared with the use of stratification alone. Adding both angular and spatial biasing to the unstratified, analog case gives an increase in efficiency of the right leakage result of 72%. This is a significant improvement compared with the use of spatial biasing alone. Finally, combining all three techniques – stratification, spatial biasing, and angular biasing – gives an increase in the efficiency of the right leakage result of 104%. For the current model, the addition of angular biasing thus provides about a 9% increase in efficiency compared with the use of stratification with spatial biasing only.

Discussion

The selections made here for spatial and angular biasing are simple and are by no means optimum. Nevertheless, even these simple schemes produce a significant increase in the efficiency of the calculation. For this example the use of angular biasing produces a 19% increase in the efficiency of calculating the right leakage for the unstratified model compared with the analog case. The addition of angular biasing also produces increases in efficiency when stratified sampling is used, but these increases are relatively small. Combining spatial and angular biasing improves the right leakage results for both the stratified and unstratified models compared with the use of either type of biasing separately. The most efficient calculation of the right leakage, which more than doubles the efficiency of the analog calculation, is obtained by combining stratification with both spatial and angular biasing.

It is apparent that the most efficient approach to a solution of this problem is to consider the leakage through only one face of the slab. Such a consideration eliminates the need for symmetry in the bias parameters with respect to the mid-plane of the slab and allows us to concentrate on choosing source particles that favor leakage through our chosen face of the slab. By seeking to maximize the efficiency of determining the right leakage and neglecting the left leakage result, we are able to obtain a relatively efficient bias scheme for the angular dependence of the source.

We could adjust the spatial biasing in a manner similar to that of the angular biasing to maximize the efficiency of calculating the right leakage. In this case we would set the probability of starting a source particle in the left-most spatial bin to a value much smaller than that of starting a source particle in the right-most bin. In the preceding calculations these spatial probabilities are equal. A calculation spatially biased for determining the leakage through only the right face of the slab would be expected to do a poor job of determining the left leakage but should significantly increase the efficiency of calculating the right leakage compared with the results presented above.

Further improvement in the efficiency of the calculation might be possible by treating only part of the geometry. Because this problem is symmetric about the mid-plane of the slab, one could place an albedo boundary coincident with this plane of symmetry. By using such a boundary one would track particles through only half of the geometry. The albedo, or reflection coefficient, of a material is defined as the ratio of the particle current coming out of the material to that going into the material; i.e., the albedo is equal to the probability that a particle striking the albedo material will be reflected from the material. By symmetry, in the present case the albedo of the hypothetical plane in the center of the slab would be unity, and such a plane would be treated as a specular reflector.

By using an albedo boundary at the center of the slab our geometry would include only half the slab and we would assume all particles striking the plane at the center are reflected with no change in weight. The new direction of travel for the reflected particles would be the mirror image of their former direction of travel. In the present case, with the albedo boundary parallel to the X-Y plane, the Z-direction cosine of the post-reflection track would be the negative of that of the pre-reflection track. No other changes would occur in the particle parameters. The source particles could be selected within the half-slab on the basis of a biasing formula that is focused on maximizing the efficiency for calculating the escape of the particles from the external surface of the half-slab. This could be done quite efficiently. The source normalization would have to be adjusted to accommodate the fact that the problem geometry now includes only half of the actual geometry.

Exercise 6.2

Statement of the Problem

2. Repeat the calculations of Example 6.5, without splitting, as well as with several different splitting parameters. Determine the efficiency of splitting when used in combination with exponential transform for this calculation.

Solution

In Example 6.5 the bias parameter ρ_o used for the exponential transform (see eqn 6.20P) is defined in subroutine 'Dist,' Table 6.33P. The split weight w_H, from eqn 6.7P, is defined in subroutine 'Col,' Table 6.34P. We can easily study the effects of changing these variables by modifying their values as defined in these subroutines. However, to facilitate such a study we have modified the routines used in solving Example 6.5 to allow the user to define these quantities as input variables. The subroutines required to perform this calculation are listed in Table 6.11.

Table 6.11. Subroutines for solving Exercise 6.2

Subroutine	Location
'Source'	Table 6.17P
'Walk'	Table 6.26P
'Bdrx'	Table 6.27P
'Bankin'	Table 6.29P
'Input'	Table 6.12
'Col'	Table 6.13
'Dist'	Table 6.14
'Stats'	Table 6.15

The modified subroutine 'Input' is shown in Table 6.12. To allow the user to specify the split weight and path stretching parameter as input variables, this subroutine has been modified to read the variables 'hifac' and 'rho.' Both of these variables are placed in common 'rho' so they are available where needed. The variable 'hifac' is the factor by which the Russian roulette survival weight w_A is multiplied in order to define w_H. As in Example 6.5, this multiplication is performed in subroutine 'Col.' The modified version of 'Col,' which uses the new variable 'hifac,' is shown in

Table 6.13. In Example 6.5 the split weight is set to two times w_A, and thus the splitting parameter of that example can be reproduced in the current model by setting 'hifac' = 2. The variable 'rho,' read in the modified subroutine 'Input,' is equal to the quantity ρ_o of eqn 6.20P. This variable is used in subroutine 'Dist' to determine the particle flight path. The modified subroutine 'Dist' is shown in Table 6.14.

Table 6.12. Subroutine 'Input' for Exercise 6.2

```
      SUBROUTINE INPUT                                                   !   1
      COMMON/IN/npart,nbatch                                             !   2
!  save hifac
      COMMON/rho/rho,hifac
     REAL(8) rho,hifac
      WRITE(*,12)                                                        !   3
   12 FORMAT(' Enter the number of particles.')                          !   4
      READ(*,*)npart                                                     !   5
      WRITE(16,*)'Number of particles = ',npart                          !   6
      WRITE(*,13)                                                        !   7
   13 FORMAT(' Enter the random number seed, an integer. ')              !   8
      READ(*,*)iseed                                                     !   9
      WRITE(16,*)' Starting random number seed is ',iseed                !  10
      CALL rndin(iseed)                                                  !  11
      nbatch=1                     ! needs to be changed if batches are to be used  12
!  read in bias factor rho and split weight factor hifac.  Baseline is hifac = 2
      WRITE(*,*)' Enter bias factor for exponential transform (<1) '
      READ(*,*)rho
      IF(rho.eq.0.) STOP
      WRITE(16,*)' Bias factor for exponential transform = ',rho
      WRITE(*,*)' Enter split weight factor (baseline = 2, no split = 0) '
      READ(*,*)hifac
      WRITE(16,*)' Split weight factor = ',hifac
      RETURN                                                             !  13
      END                                                                !  14
```

Subroutine 'Stats' as modified for this exercise is shown in Table 6.15. The particle scores and their squares are summed at entry point 'StatELp.' As in Example 6.2, the 'StatEnd' entry point calculates the expected values of the left and right leakages, and makes estimates of the standard deviations of these quantities. These results, along with certain diagnostic information from the calculations, are written to the output file.

There is no right or wrong way to select bias parameters and split weights in order to study the effect of these factors on the efficiency of the current problem. Thus, following the pattern of Table 6.35P, we select values for ρ_o of 0.8, 0.9, and 0.99 for detailed study. Without changing the Russian roulette parameters used in Example 6.5, we also select values of w_H equal to 1.3, 1.4, 1.5, 2.0, and 2.5 times w_A, as well as the case without splitting. The latter is triggered by setting 'hifac' to zero in the problem input.

Table 6.13. Subroutine 'Col' for Exercise 6.2

```
      SUBROUTINE COL                                                          !  1
      REAL(8) FLTRN,delta,WA,WL,WH,rho,hifac
      REAL(8) x,y,z,u,v,w,xo,yo,zo,uo,vo,wo,wate,age,energ                    !  3
      COMMON/PART/x,y,z,u,v,w,xo,yo,zo,uo,vo,wo,wate,age,energ,nzcur,newzn,ngroup  !  4
      REAL(8) dmfp,dtr,xsec,dcur                                              !  5
      COMMON/TRACK/dmfp,dtr,xsec,dcur                                         !  6
      REAL(8) sigt(20),c(20); ! dimensions allow up to 20 different media        7
      COMMON/GEOM/sigt,c; !sigt is total cross section, c is non-absorption prob 8
      REAL(8) colwt,rrin,rrout
      COMMON/COLWT/colwt,rrin,rrout
!  add hifac to common 'rho'
      COMMON/rho/rho,hifac
      delta=dmfp/xsec                      ! distance traveled to collision      9
      dtr=dcur+delta                       ! update total distance traveled     10
      x=xo;  y=yo;  z=zo+w*dtr   ! update position, z-direction only
      colwt=colwt+wate*(1.0d0-c(nzcur))! total weight absorbed in collisions
      wate=wate*c(nzcur)                   ! reduce wate by non-absorption probability  12
      WA=DEXP(-z*rho*10.0)*c(nzcur)/(1.0d0-rho)      ! Eqn 6.21P
      WL=WA/4.0d0
      IF(wate.LT.WL)THEN                   ! if wate small play Russian roulette
         rrin=rrin+wate                    ! total weight entering RR
         IF(wate/WA.LT.fltrn())THEN        ! if particle killed by RR
            nzcur=-1                       ! set nzcur=-1 to show particle killed
            RETURN
         ENDIF
         wate=WA                           ! particle survived RR, increase wate
         rrout=rrout+wate                  ! total weight exiting RR
      ENDIF
      CALL ISOOUT                          ! assumes isotropic scatter in lab system  20
!  use hifac to determine split weight.  Baseline case is hifac = 2
     WH=WA*hifac
     IF(hifac.GT.0.0d0.AND.wate.GT.WH)CALL BANKIN(WH,WA) ! if wate high then split
      RETURN                                                                  ! 21
      END                                                                     ! 22
```

Table 6.14. Subroutine 'Dist' for Exercise 6.2

```
      SUBROUTINE DIST                                                         !  1
      REAL(8) FLTRN,r,b,rho,hifac
      REAL(8) dmfp,dtr,xsec,dcur                                              !  3
      COMMON/TRACK/dmfp,dtr,xsec,dcur                                         !  4
      REAL(8) x,y,z,u,v,w,xo,yo,zo,uo,vo,wo,wate,age,energ                    !  5
      COMMON/PART/x,y,z,u,v,w,xo,yo,zo,uo,vo,wo,wate,age,energ,nzcur,newzn,ngroup  !  6
      REAL(8) sphcx(20),sphcy(20),sphcz(20),sphr(20),rpxmin(20),rpxmax(20),&  !  7
       rpymin(20),rpymax(20),rpzmin(20),rpzmax(20),bdin(40),bdout(40)         !  8
      LOGICAL bdtest(40), bdhit(40)                                           !  9
      COMMON/cg/sphcx, sphcy, sphcz, sphr, rpxmin, rpxmax, rpymin, rpymax, rpzmin, rpzmax, & ! 10
       bdin,bdout,nsph,nrpp,nbdy,nzones,nbz(40,41),nnext(40,40,40,2),bdtest,bdhit  ! 11
      COMMON/rho/rho,hifac
      b=1.0d0/(1.0d0-w*rho)
      r=FLTRN()                            ! pick a random number               12
      dmfp=-b*DLOG(r)                      ! transformed distance in mean free paths
      wate=wate*b*DEXP(-dmfp*(1.0d0-1.0d0/b))      ! Eqn 6.19P
      xo=x;  yo=y;  zo=z                   ! store "starting" position for track 14
      dcur=0.0D0                           ! initialize distance currently traveled  15
      DO 50 i=1,nbdy                                                          ! 16
        bdtest(i)=.FALSE.                  ! beginning a new track, no bodies yet tested 17
   50 CONTINUE                                                                ! 18
      RETURN                                                                  ! 19
      END                                                                     ! 20
```

Table 6.15. Subroutine 'Stats' for Exercise 6.2

```
SUBROUTINE Stats                                                              !  1
     REAL(8) bscore(10),bsumsq(10),cscore(10),csumsq(10),bpart(10),cpart(10)  !  2
     COMMON/STAT/bscore,bsumsq,cscore,csumsq,bpart,cpart                      !  3
     COMMON/IN/npart,nbatch                                                   !  4
     REAL(8) x,y,z,u,v,w,xo,yo,zo,uo,vo,wo,wate,age,energ                     !  5
     COMMON/PART/x,y,z,u,v,w,xo,yo,zo,uo,vo,wo,wate,age,energ,nzcur,newzn,ngroup  !  6
     REAL(8) tmp,tmp1,tmp2,var,stdev                                          !  7
     REAL(8) bank(2000,9),wtin,wtout
     COMMON/BANK/bank,ibank(2000),wtin,wtout,nsplit,nsptotal,nmax
     REAL(8) colwt,rrin,rrout
     COMMON/COLWT/colwt,rrin,rrout
 ENTRY StatOne              ! entry point to initialize arrays for complete problem  8
     bscore=0.0d0;  bsumsq=0.0d0;  cscore=0.0d0;  csumsq=0.0d0                !  9
 RETURN                                                                       ! 10
 ENTRY StatLp               ! entry point to initialize arrays for a particle   11
     bpart=0.0d0;  cpart=0.0d0                                                ! 12
 RETURN                                                                       ! 13
 ENTRY StatELp              ! entry point to store scores for a particle        14
     bscore(1)=bscore(1)+bpart(1)       ! store score                           15
     bsumsq(1)=bsumsq(1)+bpart(1)**2    ! store square for variance calculation 16
     bscore(2)=bscore(2)+bpart(2)       ! store score
     bsumsq(2)=bsumsq(2)+bpart(2)**2    ! store square for variance calculation
 RETURN                                                                       ! 19
 ENTRY StatEnd              ! entry point to calculate and print results        20
     tmp=DFLOAT(npart)                                                        ! 21
     var=bsumsq(1)/tmp-(bscore(1)/tmp)**2        ! variance of left distr
     stdev=DSQRT(var)                            ! standard deviation of left distr
     WRITE(16,*)' Left result and stdev '
     write(16,*)bscore(1)/tmp,stdev/DSQRT(tmp)   ! result and std dev of result    24
     var=bsumsq(2)/tmp-(bscore(2)/tmp)**2        ! variance of right distr
     stdev=DSQRT(var)                            ! standard deviation of right distr
     WRITE(16,*)' Right distri and stdev '
    write(16,*)bscore(2)/tmp,stdev/DSQRT(tmp)   ! result and std dev of result
     WRITE(16,*)' nmax, nsptotal, wtin, wtout '
    write(16,*)nmax,nsptotal,wtin,wtout
     tmp1=colwt+bscore(1)+bscore(2)
     tmp2=tmp1-rrout+rrin
    write(16,*)' colwt, total weight absorbed in collisions',colwt
    write(16,*)' rrin, rrout, total weight entering and ',&
   'surviving Russian roulette ',rrin, rrout
    write(16,*)' total weight absorbed and escaping tmp1',tmp1
    write(16,*)' total weight less RR kills tmp2 ', tmp2
 RETURN                                                                       ! 28
     END                                                                      ! 29
```

Using the same geometry and cross sections as those in Example 6.5, with a slab thickness of 1 cm, a total cross section of 10 cm^{-1}, and $c = 0.5$, some of the present results, obtained using 10^7 start particles in each calculation, are shown in Table 6.16. The standard deviations for the right leakages are given without results, since only these are needed to examine the efficiency. The results for the value of the leakage, not shown, are reasonable for all of the calculations.

Without splitting, both the run time and the efficiency of the calculation decrease as the value of ρ_o increases. For $\rho_o = 0.99$ the efficiency is almost a factor of 20 less than that for $\rho_o = 0.8$. Thus even more so than in Example 6.5 we find that the extreme weight changes produced by the use of highly biased exponential transform can reduce the efficiency of the calculation if used without splitting.

Table 6.16. Results for Exercise 6.2

	$\rho_o = 0.8$			$\rho_o = 0.9$			$\rho_o = 0.99$		
'hifac'	sigma x10^7	run time (sec)	ε x10^{-11}	sigma x10^7	run time (sec)	ε x10^{-11}	sigma x10^7	run time (sec)	ε x10^{-11}
0	1.297	83	7.16	1.656	65	5.61	11.80	20	0.359
1.3	1.167	90	8.16	1.165	74	9.96	3.769	23	3.06
1.4	1.173	89	8.17	1.172	73	9.97	3.788	22	3.17
1.5	1.182	88	8.13	1.181	72	9.96	3.799	22	3.15
2.0	1.217	86	7.85	1.229	69	9.60	3.989	22	2.86
2.5	1.250	84	7.62	1.273	68	9.07	4.125	22	2.67

In the five cases considered here that included splitting, the efficiency is highest for w_H having a value near 1.4 times w_A for all values of ρ_o. Also, as in Table 6.35P, for the values of ρ_o shown in Table 6.16 the efficiency is highest for $\rho_o = 0.9$. However, the results obtained thus far are insufficient to determine the choice of ρ_o that provides the maximum efficiency under the present problem constraints. Table 6.17 presents results for various values of ρ_o near 0.9 and for w_H near 1.4. These calculations show a broad maximum in the efficiency near $\rho_o = 0.88$ and $w_H = 1.4$.

Table 6.17. Results for various values of ρ_o

	w_H=1.3			w_H=1.4			w_H=1.5		
ρ_o	sigma x 10^7	run time (sec)	ε x 10^{-11}	sigma x 10^7	run time (sec)	ε x 10^{-11}	sigma x 10^7	run time (sec)	ε x 10^{-11}
0.86	1.109	83	9.80						
0.87	1.112	81	9.98	1.119	81	9.86			
0.88	1.125	79	10.00	1.131	78	10.02	1.138	78	9.90
0.89	1.140	77	9.99	1.148	76	9.98	1.156	75	9.98
0.90	1.165	74	9.96	1.172	73	9.97	1.181	72	9.96
0.91	1.198	70	9.95	1.205	70	9.84	1.214	69	9.83
0.92							1.260	65	9.69

Discussion

Optimum values for the bias parameters will depend on the specific problem being solved, and may depend on the particular computer and compiler being used to solve the problem. We have seen cases where rather extreme values of the exponential transform parameter provide significant improvement in the efficiency of a calculation, but as we will see in the following exercise, this may not always be the case. As the mesh within which the bias parameters are defined – spatial, angular, energy, etc. – is made smaller, the possibility of improvement may increase. However, there is a price paid for such improvement in the time spent by the user in defining the parameters and the computing effort required to test for and execute them. Usually a compromise is found in which some amount of detail is employed in problems that are intractable to the use of simple parameters, but the effort required for meticulous definition of detailed parameters is used in only the most intransigent problems.

Exercise 6.3

Statement of the Problem

3. Consider a cube of neutron absorbing material, 1 cm thick in each dimension, located at the center of a sphere of homogeneous scattering material of radius r. Assume the absorbing material has a zero cross section above an energy E_o and a cross section of 1 cm^{-1} below that energy. Assume the material absorbs all neutrons with an energy below E_o that experience collisions in the material; i.e., assume the material is a threshold absorber with a cutoff energy of E_o, and that c = 0 below the cutoff energy. Assume that the sphere of scattering material is composed of a homogeneous material of atomic weight A, and that neutrons of energy E_s are normally incident uniformly over the exterior surface of the sphere. Further assume that the neutrons undergo isotropic scatter in the center of mass with the atoms of this material, and that all target atoms are at rest prior to such collisions. Assume that the non-absorption probability is c and the total cross section for the scattering material is Σ_t cm^{-1}.

a. Using analog Monte Carlo, for $r = 10$, $\Sigma_t = 1$ cm^{-1}, $c = 0.9$, $E_s = 1$ MeV, and $E_o = 1$ keV, determine the probability of an incident neutron being absorbed in the neutron absorbing material.

b. Implement exponential transform and repeat the calculation of part a. How did the efficiency change? What stretching parameter provides the greatest efficiency?

c. Include survival biasing, Russian roulette, and splitting along with the exponential transform. How did these measures affect your choice of stretching parameter and the efficiency of the calculation?

Solution

There are several features of this problem that we can identify at the outset. We expect that a significant fraction of the incident neutrons will be reflected from the scattering material and will be lost to the calculation. Furthermore, because the target is both small and located deep inside a relatively large volume of scattering and absorbing material we know that, of those incident particles that are not reflected, most will be absorbed before they strike the target. Finally, of those particles that survive to reach the vicinity of the target, only those with energies below the absorption threshold of the target material can be absorbed. Even then, the absorber is but one mfp thick. Thus it is likely that many low-energy neutrons will pass through the target without experiencing an interaction. We conclude that the probability of an incident neutron being absorbed in the target will be small and that an analog calculation will require the tracking of many start particles per absorption event in the target. Appropriate variance reduction techniques should significantly improve the efficiency of this calculation.

To define the geometry for this exercise we use an RPP to describe the central cube of absorbing material, and a sphere to describe the surrounding scattering material. We use a total of three zones: the absorber, the scatterer, and an external void. We position the cube of absorber material such that it is centered at the origin with its faces perpendicular to the Cartesian axes. The sphere of scattering material is centered on the origin. The resulting geometry file 'geom.txt' is shown in Table 6.18.

Table 6.18. Geometry input for Exercise 6.3

1	- number of RPPs
-0.5 0.5 -0.5 0.5 -0.5 0.5	- body 1 - x,y,and z values for faces of RPP
1	- number of SPHs to define scatterer
0. 0. 0. 10.	- body 2 - coordinates of center and radius of SPH 1
3	- number of zones
1	- one body to define the first zone, the absorber
1	- zone 1 consists of body 1
2	- two bodies to define the second zone
2 -1	- zone 2 is inside body 2 and outside body 1
1	- one body to define the third zone
-2	- zone 3 is outside body 2

In addition to the problem geometry we must define the cross sections for the materials used in the calculation. In this exercise the cross section for zone one, the absorber, is energy dependent, being zero above the threshold energy and 1 cm^{-1} below the threshold. We choose to define the cross section for this zone for neutron energies below the threshold in the file 'xsects.txt.' We modify this zone-one cross section during particle tracking to account for the transparency of the absorber for neutrons with energies above the threshold. With this stipulation, the cross section file for Exercise 6.3 is shown in Table 6.19.

Table 6.19. Cross Section input for Exercise 6.3

0. 1.	c and Σ_t for Zone 1 for $E < E_o$
0.9 1.	c and Σ_t for Zone 2

The list of subroutines modified for this calculation is given in Table 6.20. The modified subroutine 'Source' is shown in Table 6.21. Although this problem is not spherically symmetric, there are symmetries that could be used to avoid starting source particles over the entire surface of the scattering sphere. However, we choose to distribute the source uniformly over the entire surface of the scattering sphere. This is done by finding a random location on the surface of the sphere and obtaining the direction cosines for normal incidence at that point. In subroutine 'Source' we use the library subroutine 'Isoout' to obtain a unit vector $\mathbf{\Omega}$ with direction cosines (u,v,w). With these direction cosines we determine the point at which the extension of the unit vector, when positioned at the origin, strikes the outer surface of the scattering material. This point (x,y,z) is given by eqns 3.44P, where here $x_o = y_o = z_o = 0$ and $s = 10$. However, if we select the point given by eqns 3.44P as the particle start point we must give the particle the start direction $-\mathbf{\Omega}$. Alternatively we could obtain the point at which the unit vector strikes the spherical surface in the negative direction and start the particle there in the direction $+\mathbf{\Omega}$. We use the latter technique in this exercise.

Table 6.20. Subroutines modified for Exercise 6.3

Subroutine	Location
'Source'	Table 6.21
'Isocol'	Table 6.22
'Input'	Table 6.23
'Col'	Table 6.24
'Bankin'	Table 6.25
'Bdrx'	Table 6.26
'Walk'	Table 6.27
'Dist'	Table 6.28
'Stats'	Table 6.29
'Mxsec'	Table 6.30

To calculate neutron slowing down by elastic collisions with nuclei in the scattering material we incorporate the center-of-mass, isotropic elastic scattering model of subroutine 'Isocol' shown in Table 4.9P. Because many of the variables in PFC are double precision, a few changes are required before 'Isocol' can be used with this program. These are shown in Table 6.22. 'Isocol' requires as arguments both the neutron energy E and the mass of the isotropic scatterer A. The value of A is not specified by the problem definition. Therefore we parameterize the mass of the scatterer, with A being set by the user.

Table 6.21. Subroutine 'Source' modified for Exercise 6.3

```
      SUBROUTINE SOURCE                                                          !   1
      REAL(8) x,y,z,u,v,w,xo,yo,zo,uo,vo,wo,wate,age,energ                       !   2
      COMMON/PART/x,y,z,u,v,w,xo,yo,zo,uo,vo,wo,wate,age,energ,nzcur,newzn,ngroup  !  3
      REAL(8) s
      DATA s/10.0d0/
      nzcur=2                          !  assumes source is in zone two
      CALL ISOOUT                      !  direction chosen isotropically              7
      x = -u*s; y = -v*s; z = -w*s !  particle normal into surface of sphere at radius s
      energ = 1.0d6                    ! set start energy to 1 MeV
      wate=1.0d0                       !  particle starts with a weight of one        8
      RETURN                                                                     !   9
      END                                                                        !  10
```

Table 6.22. Subroutine 'Isocol' for Exercise 6.3

```
      SUBROUTINE ISOCOL (U,V,W,A,E)
!     Subroutine ISOCOL calculates the result of elastic collisions
!     of neutrons with nucleii of mass A.  The post-collision
!     direction and energy in the laboratory system is determined
!     assuming isotropic scattering in the center of mass system.
      REAL(8) fltrn,u,v,w,e,ucm,vcm,wcm,ux,vx,wx,sr,vec,costh,sinth,phi,A,twopi
      twopi=4.0d0*DACOS(0.0d0)
      COSTH = 2.*fltrn()-1.;        PHI = fltrn()*twopi
      Wcm = (A*COSTH+1.)/DSQRT(A**2+2.*A*COSTH+1.)           !  W in CM coordinates
      SINTH = DSQRT(1.-Wcm**2)
      Ucm = SINTH*DCOS(PHI); Vcm = SINTH*DSIN(PHI)            !  U & V in CM coords
      E = E*(A*A+2.*A*COSTH+1.)/(A+1.)**2                    !  Post-collision energy
      IF (ABS(u).lt.0.9) THEN                           !  determine which transformation to use
        SR = DSQRT(1.-U*U)                              !  x-axis transformation to lab directions
        Ux = SR*UCm + U*WCm
        Vx = -U*V*UCm/SR + W*VCm/SR + V*WCm
        Wx = -W*U*UCm/SR - V*VCm/SR + W*WCm
      ELSE
        sr = DSQRT(1.0-v*v)                             !  y-axis transformation to lab directions
        ux = w*ucm/sr - u*v*vcm/sr + u*wcm
        vx = vcm*sr + v*wcm
        wx = -u*ucm/sr - v*w*vcm/sr + w*wcm
      ENDIF
      vec = DSQRT(ux*ux+vx*vx+wx*wx)                    !  normalization to correct for rounding
      u = ux/vec; v = vx/vec; w = wx/vec              !  normalized lab direction cosines
      RETURN;     END
```

Because part a of this exercise requires an analog calculation, while part b requires the use of exponential transform and part c requires the inclusion of survival biasing, Russian roulette, and splitting, we define input variables with which the user can set parameters to trigger these different variance reduction techniques. This is done in subroutine 'Input.' The resulting modified subroutine 'Input' is shown in Table 6.23. In addition to specifying the number of start particles and the start random number, the user must define the mass of the scatterer 'ascat,' the survival biasing trigger 'nsurv,' the Russian roulette survival weight 'wa,' the split weight 'wh,' and the exponential transform variable 'rho.' These quantities have been included in common 'in' for use elsewhere in the program.

Table 6.23. Subroutine 'Input' for Exercise 6.3

```
      SUBROUTINE INPUT                                                  !   1
      REAL(8) ascat,wa,wh,rho,zeta
      COMMON/IN/ascat,wa,wh,rho,zeta,npart,nbatch,nsurv
      WRITE(*,12)                                                       !   3
   12 FORMAT(' Enter the number of particles.')                         !   4
      READ(*,*)npart                                                    !   5
      WRITE(16,*)'Number of particles = ',npart                         !   6
      IF(npart.le.0)STOP
      WRITE(*,13)                                                       !   7
   13 FORMAT(' Enter the random number seed, an integer. ')             !   8
      READ(*,*)iseed                                                    !   9
      IF(iseed.LT.1)iseed=1                                             !  10
      IF(iseed.GE.2147483647)iseed=2147483646                           !  11
      WRITE(16,*)' Starting random number seed is ',iseed               !  12
      CALL rndin(iseed)                                                 !  13
      nbatch=1              ! needs to be changed if batches are to be used  14
      WRITE(*,*)' Enter the mass A of the scattering material';   READ(*,*)ascat
      WRITE(16,*)' Mass of scatterer = ',ascat
      zeta=1.0d0
      IF(ascat.GT.1.0d0)THEN                    ! use eqn 4.21P
        zeta=1.0d0+(ascat-1.0d0)**2*DLOG((ascat-1.0d0)/(ascat+1.0d0))/(2.0d0*ascat)
      ENDIF
      WRITE(*,*)' Will survival biasing be used (0/1 = no/yes)?';  READ(*,*)nsurv
      WRITE(16,*)' Survival biasing (0/1 = no/yes) ',nsurv
      WRITE(*,*)' Enter the RR survival weight,wa (0 = no RR)';     READ(*,*)wa
      WRITE(16,*)' Russian roulette weight (0 = no RR) = ',wa
      WRITE(*,*)' Enter the split weight,wh (0 = no splitting)';    READ(*,*)wh
      WRITE(16,*)' Split weight (0 = no splitting) = ',wh
      WRITE(*,*)' Enter the path stretch parameter';               READ(*,*)rho
      WRITE(16,*)' Path stretch parameter = ',rho
      RETURN                                                            !  15
      END                                                               !  16
```

In order to calculate the energy downscatter for Exercise 6.3 we replace the call to 'Isoout' in the library version of subroutine 'Col' with a call to subroutine 'Isocol.' The survival biasing, Russian roulette, and splitting is done in 'Col.' We start with the version of 'Col' shown in Table 6.28P and add tests to determine which variance reduction techniques will be used. Although the Russian roulette kill weight 'wl' could be varied, we set it equal to 'wa'/4 as was done in eqn 6.6P. A study of the effect of the ratio of 'wa'/'wl' on the efficiency of the calculation would be of interest but is not undertaken in this exercise.

The result required from this calculation is the number of neutron captures that occur in the absorbing material of zone one. All collisions are processed in subroutine 'Col' and hence we modify that subroutine to tally the desired absorption events. We use the arrays 'cscore' and 'csumsq' to record the tallies. The modified subroutine 'Col' is shown in Table 6.24. The use of the variable 'efactor,' and reference to eqn 4.18P, in the Russian roulette calculation is discussed in part c of this exercise.

Table 6.24. Subroutine 'Col' used in Exercise 6.3

```
      SUBROUTINE COL                                                          !      1
      REAL(8) FLTRN,delta,wavg,rad,whigh,efactor,avgwt,tmp
      REAL(8) x,y,z,u,v,w,xo,yo,zo,uo,vo,wo,wate,age,energ                    !      3
      COMMON/PART/x,y,z,u,v,w,xo,yo,zo,uo,vo,wo,wate,age,energ,nzcur,newzn,ngroup  !  4
      REAL(8) dmfp,dtr,xsec,dcur                                              !      5
      COMMON/TRACK/dmfp,dtr,xsec,dcur                                         !      6
      REAL(8) sigt(20),c(20);  ! dimensions allow up to 20 different media          7
      COMMON/GEOM/sigt,c; !sigt is total cross section, c is non-absorption prob    8
      REAL(8) rrin,rrout
      COMMON/RR/rrin,rrout
      REAL(8) ascat,wa,wh,rho,zeta,wl
      COMMON/IN/ascat,wa,wh,rho,zeta,npart,nbatch,nsurv
      REAL(8) cscore,csumsq,bscore,bsumsq
      COMMON/STAT/cscore,csumsq,bscore,bsumsq
      delta=dmfp/xsec                    ! distance traveled to collision           9
      dtr=dcur+delta                     ! update total distance traveled          10
      x=xo+u*dtr;  y=yo+v*dtr;  z=zo+w*dtr    ! update position                    11
      Collision_location: SELECT CASE(nzcur)
        CASE(1)                          ! collision in absorber
          cscore=cscore+wate             ! score real collision (an absorption)
          csumsq=csumsq+wate**2
          nzcur=-1
          RETURN
        CASE(2)                          ! collision in scatterer
          IF(nsurv.LE.0)THEN             ! if analog scattering
            IF(fltrn().GT.c(nzcur))THEN      ! if absorbed during scattering event
              nzcur=-1
              RETURN
            ENDIF
          ELSE                           ! survival biasing
            wate=wate*c(nzcur)           ! reduce wate by non-absorption probability
          ENDIF
      END SELECT Collision_location
      IF(wa.GT.0.) THEN                  ! if Russian roulette on
        rad=DSQRT(x**2+y**2+z**2)
        avgwt=1.0d0
        IF(nsurv.GT.0)THEN
          tmp=energ; IF(tmp.LT.10.0d0)tmp=10.0d0
          efactor=DLOG(1.0d6/tmp)/zeta        ! eqn 4.18P
          avgwt=c(nzcur)**(efactor)
        ENDIF
        wavg=wa*avgwt*DEXP(-(10.0d0-rad)*rho)/(1.0d0-rho)    ! eqn 6.21P modified
        wl=wavg/4.0d0;  whigh=wh*wavg
        IF(wate.LT.wl)THEN               ! if wate small play Russian roulette
          rrin=rrin+wate                 ! total weight entering RR
          IF(wate/wavg.LT.fltrn())THEN    ! if particle killed by RR
              nzcur=-1                   ! set nzcur=-1 to show particle killed
             RETURN
          ENDIF
          wate=wavg                       ! particle survived RR, increase wate
          rrout=rrout+wate               ! total weight exiting RR
        ENDIF
      ENDIF
      CALL ISOCOL(u,v,w,ascat,energ)    ! assumes isotropic scatter in CM system
      IF(wa.GT.0.0d0.AND.wh.GT.0.0d0.AND.wate.GT.whigh)CALL BANKIN(whigh,wavg) !split on
      RETURN                                                                  !     21
      END                                                                     !     22
```

To incorporate splitting we add subroutine 'Bankin' to PFC, as described in Example 6.4. In order to do this, modifications are needed to bank the energy, as well as to store and retrieve the direction cosines of the particles banked. The modified subroutine 'Bankin' is shown in Table 6.25.

Table 6.25. Subroutine 'Bankin' for Exercise 6.3

```
SUBROUTINE BANKIN(WH,WA)
REAL(8) wh,wa
REAL(8) x,y,z,u,v,w,xo,yo,zo,uo,vo,wo,wate,age,energ
COMMON/PART/x,y,z,u,v,w,xo,yo,zo,uo,vo,wo,wate,age,energ,nzcur,newzn,ngroup
REAL(8) bank(2000,10),wtin,wtout
COMMON/BANK/bank,ibank(2000),wtin,wtout,nsplit,nsptotal,nmax
IF(nsplit.LT.2000)THEN
  nsplit=nsplit+1                    ! add to bank if not full
ELSE
  RETURN
ENDIF
IF(nsplit.GT.nmax)nmax=nsplit
nsptotal=nsptotal+1
wtin=wtin+wate-WA
bank(nsplit,1)=x;  bank(nsplit,2)=y;  bank(nsplit,3)=z          ! bank position
bank(nsplit,4)=u;  bank(nsplit,5)=v;  bank(nsplit,6)=w        ! bank direction
bank(nsplit,7)=wate-WA;  bank(nsplit,8)=WH;  bank(nsplit,9)=WA ! bank weights
bank(nsplit,10)=energ
wate=WA                                  ! set weight of remaining particle
ibank(nsplit)=nzcur
RETURN
ENTRY BANKOUT                            ! entry to fetch particle
x=bank(nsplit,1);  y=bank(nsplit,2);  z=bank(nsplit,3)        ! get position
u = bank(nsplit,4); v=bank(nsplit,5); w=bank(nsplit,6)
energ=bank(nsplit,10)
nzcur=ibank(nsplit)
IF(bank(nsplit,7).LT.bank(nsplit,8))THEN  ! see if remaining weight<WH
  wate=bank(nsplit,7)                     ! if so retrieve all weight
  nsplit=nsplit-1                         ! particle weight no longer in bank
ELSE
  wate=bank(nsplit,9)                     ! retrieve WA of banked weight
  bank(nsplit,7)=bank(nsplit,7)-bank(nsplit,9)      ! reduce banked weight
ENDIF
wtout=wtout+wate
RETURN
END
```

Because we anticipate that many of the incident neutrons will be reflected from the sphere of scattering material, we count such reflections in the variable 'bscore.' To do this we modify subroutine 'Bdrx' to increment 'bscore' when a particle escapes from the geometry. The modified 'Bdrx' is shown in Table 6.26. The version of 'Walk' shown in Table 6.26P has been modified to reflect the different common statements used in this exercise. This revised version of 'Walk' is shown in Table 6.27.

To accommodate exponential transform, additional modifications beyond those used for Example 6.5 must be made to subroutine 'Dist.' In the current exercise the path-stretching variable 'rho' is stored in common 'in' and this common statement must now be included in 'Dist.' Unlike the stretching in the positive-Z direction done in Example 6.5, here we want to stretch toward the origin. Therefore let us define μ as the negative of the scalar product of the current flight path direction, $\mathbf{\Omega}$, and the radial vector from the origin to the collision point, $\mathbf{\Omega}_c$,

$$\mu = -\Omega \bullet \Omega_c = -\frac{ux + vy + wz}{R}$$

where

$$R = \sqrt{x^2 + y^2 + z^2}$$

Table 6.26. Subroutine 'Bdrx' for Exercise 6.3

```
SUBROUTINE BDRX                                                              !  1
REAL(8) delta                                                                !  2
REAL(8) sphcx(20),sphcy(20),sphcz(20),sphr(20),rpxmin(20),rpxmax(20),&       !  3
 rpymin(20),rpymax(20),rpzmin(20),rpzmax(20),bdin(40),bdout(40)              !  4
LOGICAL bdtest(40), bdhit(40)                                                !  5
COMMON/cg/sphcx,sphcy,sphcz,sphr,rpxmin,rpxmax,rpymin,rpymax,rpzmin,rpzmax,& !  6
bdin,bdout,nsph,nrpp,nbdy,nzones,nbz(40,41),nnext(40,40,40,2),bdtest,bdhit   !  7
REAL(8) x,y,z,u,v,w,xo,yo,zo,uo,vo,wo,wate,age,energ                         !  8
COMMON/PART/x,y,z,u,v,w,xo,yo,zo,uo,vo,wo,wate,age,energ,nzcur,newzn,ngroup  !  9
REAL(8) dmfp,dtr,xsec,dcur                                                   ! 10
COMMON/TRACK/dmfp,dtr,xsec,dcur                                              ! 11
REAL(8) cscore,csumsq,bscore,bsumsq
COMMON/STAT/cscore,csumsq,bscore,bsumsq
delta=dtr-dcur                 ! delta--distance traveled to reach boundary    12
dcur=dtr                       ! update current distance traveled              13
dmfp=dmfp-delta*xsec           ! subtract current distance in mfp from dmfp    14
x=x+u*delta; y=y+v*delta;  z=z+w*delta    ! update position                    15
nzcur=newzn                    ! change identifier of current zone             16
IF(nzcur.LT.nzones)RETURN      ! if not at outer zone then return              17
nzcur=-1                       ! set nzcur=-1 if in outer zone (for escape)    18
bscore=bscore+wate
bsumsq=bsumsq+wate**2
RETURN                                                                       ! 19
END                                                                          ! 20
```

For maximum efficiency we anticipate that the stretching toward the origin should be greatest at the edge of the sphere, and should decrease as the particles approach the origin. Thus we make the variable ρ in eqn 6.16P proportional to the radius at which a collision occurs, and normalize it such that it is unity at the outer surface of the scattering sphere. Thus for a sphere of radius $R_o = 10$ we have

$$\rho = \frac{\mu R}{R_o} = -\frac{ux + vy + wz}{10} \tag{6.1}$$

This definition of ρ must be substituted into eqn 6.16P and used in the definition of the variable 'b' in subroutine 'Dist.' Starting with the version of 'Dist' from Table 6.33P, the subroutine as modified for the present calculation is shown in Table 6.28.

Table 6.27. Subroutine 'Walk' for Exercise 6.3

```
      SUBROUTINE WALK                                                    !    1
      REAL(8) ascat,wa,wh,rho,zeta
      COMMON/IN/ascat,wa,wh,rho,zeta,npart,nbatch,nsurv
      REAL(8) dmfp,dtr,xsec,dcur                                         !    3
      COMMON/TRACK/dmfp,dtr,xsec,dcur                                    !    4
      REAL(8) x,y,z,u,v,w,xo,yo,zo,uo,vo,wo,wate,age,energ               !    5
      COMMON/PART/x,y,z,u,v,w,xo,yo,zo,uo,vo,wo,wate,age,energ,nzcur,newzn,ngroup  !  6
      REAL(8) bank(2000,10),wtin,wtout
      COMMON/BANK/bank,ibank(2000),wtin,wtout,nsplit,nsptotal,nmax
      REAL(8) colwt,rrin,rrout
      COMMON/COLWT/colwt,rrin,rrout
      nsptotal=0;  nmax=0; wtin=0.0d0; wtout=0.0d0;colwt=0.0d0
      rrin=0.0d0;  rrout=0.0d0
Loop_Over_Particles: DO i=1,npart     ! loop over number of particles              7
      CALL StatLp                     ! initialize statistics for each particle    8
      CALL SOURCE                     ! get source parameters for particle         9
      nsplit=0
  100 CONTINUE
  Loop_for_Collisions: DO             ! to find collision sites                   10
      CALL DIST                       ! get distance of travel in mfp             11
    Loop_Track:  DO                   ! track particle to collision point         12
      CALL HIT                        ! get distance to boundary of current zone  13
      CALL MXSEC                      ! get total cross section for current zone  14
      IF((dtr-dcur)*xsec.GE.dmfp) EXIT Loop_Track  ! collision before boundary    15
      CALL BDRX                       ! process boundary crossing                 16
      IF(nzcur.EQ.-1) EXIT Loop_for_Collisions  ! particle killed e.g., by escaping 17
    END DO Loop_Track                                                    !   18
      CALL COL                        ! call col to process collision             19
      IF(nzcur.EQ.-1) EXIT Loop_for_Collisions  ! particle killed by collision    20
  END DO Loop_for_Collisions                                             !   21
      IF(nsplit.GT.0)THEN
        CALL BANKOUT
        GOTO 100
      ENDIF
      CALL StatELp                    ! process statistics when particle killed   22
END DO Loop_Over_Particles                                               !   23
      RETURN                                                             !   24
      END                                                                !   25
```

To define the tally for the absorptions in zone one, and to estimate the uncertainties in the result, we must modify subroutine 'Stats.' We start with the version of 'Stats' shown in Table 6.30P. In the present version of the routine the standard deviation of the tally array 'cscore' is calculated in the usual manner and the results are written to the output file. The probability of incident particles being reflected, plus the number of split particles, Russian roulette tests, and Russian roulette kills are also written to the output file. The modified version of 'Stats' is shown in Table 6.29.

We must now implement the absorption threshold for the material in zone one. There are several ways we can do this. For example, for high-energy particles we could allow only virtual collisions to occur in the absorber. In such collisions the direction and energy of the particle are not changed. Use of such virtual collisions takes advantage of the fact, as discussed in Example 3.4, that the flight path of a particle can be reselected at any point along the particle's track without changing the results of the calculation. Alternatively, for high energy neutrons we could specify a zero total cross section in zone one whereby neutrons above the threshold cannot experience a collision in the zone. Although we know from eqn 3.43P that the distance to the next collision is inversely proportional to the total cross section of the material through which a particle is traveling, the coding in PFC does not calculate $1/\Sigma_t$ directly and we are free to set Σ_t to zero. For this exercise we use the latter method. The modified subroutine 'Mxsec' is shown in Table 6.30.

Table 6.28. Subroutine 'Dist' for Exercise 6.3

```
      SUBROUTINE DIST                                                          !  1
      REAL(8) FLTRN,r,b,rhofac
      REAL(8) dmfp,dtr,xsec,dcur                                               !  3
      COMMON/TRACK/dmfp,dtr,xsec,dcur                                          !  4
      REAL(8) x,y,z,u,v,w,xo,yo,zo,uo,vo,wo,wate,age,energ                     !  5
      COMMON/PART/x,y,z,u,v,w,xo,yo,zo,uo,vo,wo,wate,age,energ,nzcur,newzn,ngroup  !  6
      REAL(8) sphcx(20),sphcy(20),sphcz(20),sphr(20),rpxmin(20),rpxmax(20),&   !  7
       rpymin(20),rpymax(20),rpzmin(20),rpzmax(20),bdin(40),bdout(40)          !  8
      LOGICAL bdtest(40), bdhit(40)                                            !  9
      COMMON/cg/sphcx,sphcy,sphcz,sphr,rpxmin,rpxmax,rpymin,rpymax,rpzmin,rpzmax,&  ! 10
       bdin,bdout,nsph,nrpp,nbdy,nzones,nbz(40,41),nnext(40,40,40,2),bdtest,bdhit   ! 11
      REAL(8) ascat,wa,wh,rho,zeta
      COMMON/IN/ascat,wa,wh,rho,zeta,npart,nbatch,nsurv
      IF(rho.gt.0.) THEN
        rhofac = -rho*(x*u+y*v+z*w)/10.0d0
        b=1.0d0/(1.0d0-rhofac)
        r=FLTRN()                          ! pick a random number
        dmfp=-b*DLOG(r)                    ! transformed distance in mean free paths
        wate = wate*b*DEXP(-dmfp*(1.0d0-1.0d0/b))
      ELSE
        dmfp = -DLOG(fltrn())
      ENDIF
      xo=x;  yo=y;  zo=z                   ! store "starting" position for track       14
      dcur=0.0D0                           ! initialize distance currently traveled     15
      DO 50 i=1,nbdy                                                           ! 16
        bdtest(i)=.FALSE.                  ! beginning a new track, no bodies yet tested 17
   50 CONTINUE                                                                 ! 18
      RETURN                                                                   ! 19
      END                                                                      ! 20
```

Results Part a

We are now ready to execute PFC for this exercise. For part a we set all of the variance reduction parameters to zero in order to run an analog calculation. In this case, using 10^7 source particles and a start random number of one, we obtain the probabilities of reflection and of absorption in the target that are shown in Table 6.31. The results for A = 1 appear to be acceptable; however, based on the standard deviations shown, those for A = 6 are marginal and those for A = 12 are highly questionable. In the latter case, our calculation estimates the probability of a source particle being absorbed in the target to be less than 4 parts in 10^6. Since we started 10^7

particles in this calculation, we apparently obtained only 36 absorptions. The standard deviation of the A = 12 result shown in Table 6.31 reflects this small sample size.

Table 6.29. Subroutine 'Stats' for Exercise 6.3

```
SUBROUTINE Stats                                                              !    1
      REAL(8) cscore,csumsq,bscore,bsumsq
      COMMON/STAT/cscore,csumsq,bscore,bsumsq
      REAL(8) ascat,wa,wh,rho,zeta
      COMMON/IN/ascat,wa,wh,rho,zeta,npart,nbatch,nsurv
      REAL(8) x,y,z,u,v,w,xo,yo,zo,uo,vo,wo,wate,age,energ                   !    5
      COMMON/PART/x,y,z,u,v,w,xo,yo,zo,uo,vo,wo,wate,age,energ,nzcur,newzn,ngroup  !  6
      REAL(8) tmp,tmp1,tmp2,var,stdev                                         !    7
      REAL(8) bank(2000,10),wtin,wtout
      COMMON/BANK/bank,ibank(2000),wtin,wtout,nsplit,nsptotal,nmax
      REAL(8) rrin,rrout
      COMMON/RR/rrin,rrout
 ENTRY StatOne               ! entry point to initialize arrays for complete problem  8
      bscore=0.0d0;  bsumsq=0.0d0;  cscore=0.0d0;  csumsq=0.0d0              !    9
 RETURN                                                                       !   10
 ENTRY StatLp                ! entry point to initialize arrays for a particle   11
 RETURN                                                                       !   13
 ENTRY StatELp               ! entry point to store scores for a particle       14
 RETURN                                                                       !   19
 ENTRY StatEnd               ! entry point to calculate and print results       20
      tmp=DFLOAT(npart)                                                       !   21
      var=bsumsq/tmp-(bscore/tmp)**2        ! variance of bscore distribution
      stdev=DSQRT(var)                            ! standard deviation of bscore distr23
      WRITE(16,*)' Escape probability = ',bscore/tmp,stdev/DSQRT(tmp)
      var=csumsq/tmp-(cscore/tmp)**2        ! variance of cscore distribution
      stdev=DSQRT(var)                            ! standard deviation of cscore distr26
      WRITE(16,*)' Absorption probability = ',cscore/tmp,stdev/DSQRT(tmp)
! RR and split statistics
      WRITE(16,*)' RRin, RRout ',rrin, rrout
      WRITE(16,*)' Number of splits, Bankwtin, Bankwtout ',nsptotal,wtin, wtout
 RETURN                                                                       !   28
      END                                                                     !   29
```

Table 6.30. Subroutine 'Mxsec' for Exercise 6.3

```
      SUBROUTINE MXSEC                                                        !    1
      REAL(8) dmfp,dtr,xsec,dcur                                              !    2
      COMMON/TRACK/dmfp,dtr,xsec,dcur                                         !    3
      REAL(8) x,y,z,u,v,w,xo,yo,zo,uo,vo,wo,wate,age,energ                    !    4
      COMMON/PART/x,y,z,u,v,w,xo,yo,zo,uo,vo,wo,wate,age,energ,nzcur,newzn,ngroup  !  5
      REAL(8) sigt(20),c(20)                                                  !    6
      COMMON/GEOM/sigt,c  ! sigt is total cross section, c is non-absorption prob    7
      xsec=sigt(nzcur)                                                        !    8
      if(nzcur.eq.1.and.energ.gt.1000.)xsec = 0.0
      RETURN                                                                  !    9
      END                                                                     !   10
```

Table 6.31. Results for Exercise 6.3 Analog Calculation

A	Run time (sec)	Reflection	stdev	Abs in target	stdev	ε
1	150	0.3074	1.459-4	4.072-4	6.380-6	1.64+8
6	124	0.4557	1.575-4	3.430-5	1.852-6	2.35+9
12	122	0.4656	1.577-4	3.600-6	6.000-7	2.28+10
56	120	0.4730	1.579-4	0	—	—

The present analog calculation for A = 56, which uses 10^7 start particles, produces no absorptions in the target. We conclude, therefore, that the probability of an absorption in the target is probably less than one in 10^7 for this case. We examine the case of A = 56 in greater detail below and, as we shall see, the probability of absorption in the target per start particle in this case is in fact much less than 10^{-7}.

We see from Table 6.31 that the probability of an incident neutron being reflected from the sphere varies with the mass of the scattering material. This is a result of the fact that elastic scattering of neutrons on low-mass targets results in an angular distribution for the post-collision neutrons that is highly forward peaked in the laboratory system. This angular distribution becomes increasingly isotropic as the mass of the target nuclei increases. Therefore the mean number of collisions required to reverse the direction of a high-energy neutron by isotropic scattering in the center of mass decreases with increasing target mass. This is discussed in Example 4.2. Thus, for the present exercise, the probability of an incident neutron being reflected from the scattering sphere increases with increasing target mass, for the same non-absorption probability c.

The probability of an incident neutron being absorbed in the target material in this exercise decreases with increasing mass of scattering material. This is a consequence of both the increased reflection of incident neutrons and the fact that the number of collisions required to reduce the energy of an incident neutron to or below the absorption threshold of the target by elastic scattering increases with the mass of the nuclei of the scattering material. The latter effect is discussed in Example 4.1. Because in this exercise the non-absorption probability is constant in the scattering material, the probability of a neutron surviving enough collisions to have its energy decreased by a given amount decreases with increasing target mass.

Results Part b

As expected, the analog solution to this exercise shows that only a small fraction of the incident neutrons produce absorptions in the target. Thus the use of an appropriate exponential transform procedure to bias the calculation should increase the efficiency. In setting up this problem we include simple angle- and space-dependent path stretching in subroutine 'Dist,' as shown in Table 6.28. We know that the selection of this angular-dependent path stretching parameter, based on eqn 6.20P, will provide maximum stretching when Ω and Ω' are parallel, where here Ω' is a unit vector pointing toward the origin. Thus, according to the current version of subroutine 'Dist,' for a given value of ρ_o, the initial flight path of all source particles will undergo maximum stretching. Because of this biasing, a large value of ρ_o could cause the first flight path of a particle to pass completely through the absorbing material, or even completely through the scattering sphere, and result in a lower efficiency than a small stretching parameter. We anticipate that some intermediate value of ρ_o will give the maximum improvement in calculational efficiency.

The results obtained from running the present modified version of PFC with 10^7 start particles, for A = 1, with path stretching parameter values ρ_o from 0.1 to 0.9, are shown in Table 6.32. The result for $\rho_o = 0$ is that from the first row of Table 6.31. The efficiency with which the absorption in the target is calculated, as determined by eqn 6.2P, is shown in the last column of the table. It appears that a small amount of path stretching toward the center of the sphere reduces the standard deviation of the absorption probability, although when the resulting increased run time is taken into account the efficiency is decreased. As ρ_o is increased above a value of 0.2 the standard deviation and the efficiency change only slightly, but the latter appears to reach a maximum near ρ_o = 0.215. Further increases in ρ_o decrease the efficiency of the calculation.

Table 6.32. Results for Exercise 6.3, Exponential Transform, A = 1

ρ_o	Run time (sec)	Refl	stdev	Abs	stdev	ε
0.	150	0.3074	1.459-4	4.072-4	6.380-6	1.64+8
0.1	182	0.3073	1.544-4	4.054-4	5.955-6	1.53+8
0.2	185	0.3070	1.691-4	4.063-4	5.770-6	1.62+8
0.21	184	0.3070	1.709-4	4.010-4	5.715-6	1.66+8
0.215	184	0.3070	1.719-4	3.998-4	5.705-6	1.67+8
0.22	184	0.3071	1.729-4	4.017-4	5.721-6	1.66+8
0.3	189	0.3074	1.918-4	4.090-4	5.798-6	1.57+8
0.5	197	0.3075	2.774-4	4.100-4	7.222-6	9.73+7
0.9	183	0.3081	2.141-3	3.676-4	2.222-5	1.11+7

The results shown in Table 6.32 provide only an approximation to the optimum path stretching for this problem and apply only to the particular path stretching model given by eqn 6.1. Furthermore, as discussed under Exercise 6.1, in the present case, where we are calculating efficiencies and are attempting to use several significant digits in the variance results in order to compare the efficiencies of different calculations, we should seek to obtain fsd values much less than 0.1. The absorption results in Table 6.32 have fsd values of approximately 0.015, which indicates that the estimated variances are acceptable. The location of the maximum in the efficiency is still subject to the error introduced by the uncertainties in the run times, however. Both the run time and the location of the maximum in the efficiency could change if the same problem were executed on a different computer or if a different Fortran compiler or compiler option were used.

Despite the fact that some improvement in the efficiency with which this calculation is performed is apparently obtained by including exponential transform, the increase in efficiency compared with the analog case is insignificant. Furthermore, unless a relatively good value for ρ_o is used, implementation of path stretching decreases the efficiency compared with the analog result. Therefore, the value of employing exponential transform by itself appears to be marginal in this problem.

Results Part c

We now want to determine whether survival biasing, Russian roulette, and splitting, used in combination with exponential transform, will improve the efficiency of our calculation. These techniques can be incorporated in a variety of ways, such as introducing one or two at a time, or introducing all at once, while choosing various values for the required parameters. We begin by introducing survival biasing and Russian roulette, and add splitting later.

In Example 4.1 we examine the slowing down of neutrons by elastic collisions in a material of fixed atomic mass. In that example we make use of two analytical estimates for the average number of collisions N required to reduce the energy of a neutron from a given initial energy E_o to some lower energy E_f, and we compare these estimates with our Monte Carlo results. The calculations show that the geometric average number of collisions, given by eqn 4.18P, is more accurate for estimating the true number of collisions required to slow a neutron to a given energy than the linear average of eqn 4.17P. For the present exercise, then, we choose the Russian roulette survival weights using eqn 4.18P.

Since the non-absorption probability c is equal to the fraction of the incident particle weight that will emerge from a collision when survival biasing is used, the weight emerging from the Nth collision experienced by a neutron is

$$w_N = w_o c^N \tag{6.2}$$

where w_o is the initial weight of the particle. We thus specify a value for the Russian roulette survival weight w_A based on analog scattering, and when survival biasing is included we multiply this weight by w_N to determine the survival weight for a particle of energy E_f where, by eqn 4.18P, the value of N is estimated using

$$N = \left[\left[\frac{1}{\zeta}\ln\frac{E_o}{E_f}\right]\right] \tag{6.3}$$

Here E_0 is the energy of the source neutrons and ζ is the average logarithmic energy decrement per collision given by eqn 4.21P.

The version of subroutine 'Col' shown in Table 6.24 uses eqns 6.2 and 6.3 to estimate the average weight that a neutron should have as a function of its current energy. The value of ζ is calculated in subroutine 'Input.' In order to avoid tracking particles that have extremely small weights, we need to select some reasonable lower bound for the neutron weights. This bound should be consistent with the actual weights of the neutrons that reach the threshold of the absorber, but should allow us to avoid tracking particles that make negligible contributions to the final tally. We know that when survival biasing is used the neutrons continue to lose weight as their energy decreases below the absorber threshold, and therefore we know that scores from neutrons with energies near the threshold will be larger, on the average, than scores from neutrons with energies below the threshold. Therefore we should be able to find a cutoff energy below which contributions to the final tally can be neglected. Although this lower weight limit will be dependent on the mass of the target nuclei, a series of calculations for A = 12 shows that a reasonable value to use in this model is the average particle weight predicted by eqns 6.2 and 6.3 for a neutron with an energy of 10 eV. We use this cutoff in the calculations presented here.

To examine the effect of survival biasing and Russian roulette on the efficiency of the calculation without exponential transform, we initially restrict our analysis to the relatively simple case of A = 1. Running PFC with the modified subroutines listed in Table 6.20 including survival biasing, with several Russian roulette survival weights, and with 10^7 start particles, gives the results for the probability of absorption in the target shown in Table 6.33. In order to determine the efficiencies of the various calculations with reasonable accuracy, relatively long run times are used. A value of w_A = 1.0 produces the most efficient calculation, although the efficiency appears to be relatively insensitive to this parameter.

Table 6.33. Results of Exercise 6.3 with Survival Biasing and Russian Roulette for A=1

w_A	Run Time (sec)	Absorption	Std Dev	ε
0.9	405	4.165-4	3.312-6	2.251+8
1.0	400	4.105-4	3.301-6	2.294+8
1.1	395	4.125-4	3.324-6	2.291+8
2.0	357	4.102-4	3.727-6	2.017+8

Using w_A = 1.0 and applying survival biasing and Russian roulette to the cases for the values of A shown in Table 6.31 gives the results shown in Table 6.34. The improvement in efficiency obtained by using these techniques for the case of A = 1, compared with the analog calculation, is about 40%, while the improvement obtained for larger values of A is much greater than this. In the case of A = 56, an analog run would need almost 10^{16} particles to obtain the same results as the case shown here using 10^7 particles with survival biasing and Russian roulette. This means that the improvement in efficiency is about a factor of 10^8. For large values of the scattering target mass the loss of particles through absorption in the downscattering process dominates the transport calculation. Therefore, for high-mass scatterers the use of survival biasing and Russian roulette, with parameters chosen to account for the absorption during downscattering, is essential in order to obtain a solution.

Table 6.34. Results for Exercise 6.3. Comparison of Analog vs. Survival Biasing and Russian Roulette

	Analog				Survival Biasing and Russian Roulette			
A	Run Time (s)	Absorp	Std Dev	ε	Run Time (s)	Absorp	Std Dev	ε
1	150	4.072-4	6.380-6	1.64+8	400	4.105-4	3.301-6	2.29+8
6	124	3.430-5	1.852-6	2.35+9	409	3.808-5	4.446-7	1.24+10
12	122	3.600-6	6.000-7	2.28+10	450	3.125-6	4.757-8	9.82+11
56	120	0	---	---	475	1.221-14	2.191-15	4.38+26

Even for small values of A, in which the loss of particles by absorption prior to reaching the threshold energy is relatively unimportant, the small size of the absorber introduces a significant level of difficulty into this problem. An improvement in efficiency over that of the analog case is highly desirable. We expect that, with additional effort, we can obtain greater improvement in the efficiency of our calculation than that shown in Table 6.34. Thus let us examine the combination of exponential transform with survival biasing, Russian roulette, and splitting.

Table 6.35 shows the results of calculations using A = 1 with various values of ρ_o while incorporating both Russian roulette and splitting. Values of $w_A = 1.0$ and $w_H = 2w_A$ have been used. That is, w_H is chosen such that it tracks the energy-dependent values of Russian roulette survival weight that are obtained as explained above. For the cases shown in Table 6.35 the highest efficiency, which occurs for $\rho_o = 0.92$, is about a factor of 29 greater than the analog efficiency.

Table 6.35. Results of Exercise 6.3 with Survival Biasing, Russian Roulette, Splitting, and Exponential Transform, A = 1

ρ_o	Run Time (s)	Absorp	Std Dev	ε
0 (analog)	150	4.072-4	6.380-6	1.638+8
0.5	454	4.139-4	1.266-6	1.374+9
0.8	873	4.130-4	5.577-7	3.683+9
0.9	81	4.125-4	1.647-6	4.551+9
0.92	74	4.078-4	1.689-6	4.737+9
0.95	58	4.173-4	1.928-6	4.638+9

It is apparent that, at least for A = 1, the effect of particle splitting on the efficiency of this problem is significant. Let us now parameterize w_H to see whether the effect of splitting can be further improved. Still limiting ourselves to the case of A = 1, the results for exponential transform with ρ_o = 0.92 and w_A= 1.0 are shown in Table 6.36 for various values of w_H. The best results obtained are for w_H = 1.2. For this case the efficiency is approximately 32 times that of the analog calculation.

Table 6.36. Results of Exercise 6.3 with Survival Biasing, Russian Roulette, Splitting, and Exponential Transform, A = 1

w_H	Run Time (s)	Absorp	Std Dev	ε
1.0	107	4.110-4	1.356-6	5.083+9
1.2	99	4.126-4	1.386-6	5.258+9
1.4	92	4.202-4	1.480-6	4.962+9
1.6	83	4.165-4	1.553-6	4.996+9
1.8	78	4.097-4	1.622-6	4.873+9
2.0	74	4.078-4	1.689-6	4.737+9
2.2	71	4.137-4	1.770-6	4.496+9

Using the same values for path stretching and the Russian roulette survival weight, ρ_o = 0.92 and w_A= 1.0, results obtained for various values of w_H with A = 12, both with and without the use of survival biasing, are shown in Table 6.37. The most efficient result obtained without survival biasing that is shown in this table, which is for w_H = 1.4, gives an efficiency about 72 times that of the analog calculation of Table 6.34. When survival biasing is added a further improvement of a factor of about 17.8 is achieved.

The present results for A = 12 show that both exponential transform and a combination of survival biasing, Russian roulette, and splitting, provide significant improvements in efficiency over the analog case for this problem. The exponential transform technique, which is intended to compensate for the small detector, is slightly more effective than the techniques that address the changes in particle weight during scattering. As should be expected, the best results are obtained when all of the techniques are used together, as shown in the last row of Table 6.37. In this case the efficiency of the calculation is about 1282 times greater than that of the analog case.

Table 6.37. Results of Exercise 6.3 for A=12, w_A = 1.0, ρ_o = 0.92

w_H	Survival Bias, 1 = yes	Run Time (sec)	Absorp	Std Dev	ε
1.2	0	436	3.173-6	3.842-8	1.553+12
1.4	0	409	3.189-6	3.864-8	1.638+12
1.6	0	382	3.278-6	4.198-8	1.485+12
1.8	0	364	3.143-6	4.290-8	1.493+12
2.0	0	345	3.157-6	4.455-8	1.160+12
1.4	1	410	3.170-6	9.135-9	2.923+13

The results obtained for this problem from running the current version of PFC for several values of w_H, using A = 56, ρ_o = 0.92 and w_A= 1.0, are shown in Table 6.38. In all of these cases survival biasing is necessary in order to obtain any results at all. Even with survival biasing, run times of about 1.5 hours are required in order to get good statistics from which efficiency comparisons can be made. The highest efficiency for this case is obtained for w_H = 1.2. Thus the addition of exponential transform and splitting to the survival biasing and Russian roulette calculation shown in Table 6.34 provides an additional improvement of about a factor of 142 in the efficiency.

Table 6.38. Results of Exercise 6.3 for A=56, w_A = 1.0, ρ_o = 0.92

w_H	Run Time (s)	Absorp	Std Dev	ε
1.0	5824	9.919-15	5.412-17	5.863+28
1.2	5453	9.762-15	5.428-17	6.224+28
1.4	5055	9.702-15	5.728-17	6.029+28

Discussion

We have by no means exhausted the space of Russian roulette, splitting, and exponential transform biasing parameters, along with possible spatial and energy variations in the same, that are possible for this exercise. However, the selection of a reasonable set of parameters has shown that significant improvements in efficiency can be obtained with these techniques.

Russian roulette and splitting are important supporting techniques when exponential transform is used. Furthermore, survival biasing can make significant improvements in efficiency when used in combination with reasonable Russian roulette and splitting parameters. Exponential transform can help ensure that important geometric regions in the problem are adequately sampled, while survival biasing can do the same for particle energies that require a number of collision events to reach.

This exercise presents a somewhat complex and difficult, but not unrealistic, problem. That is, it is representative of Monte Carlo particle transport problems that require the practitioner to estimate particle fluxes or other quantities in small regions that are located at significant distances (in mfp) from the source. Problems of this type frequently require considerable investment of time to identify variance reduction techniques, and their specific parameter values as a function of the problem variables such as space, angle, and energy, that will provide meaningful improvements in efficiency.

Chapter 7

Monte Carlo Detectors

Exercise 7.1

Statement of the problem

1. Repeat Example 7.1, but place small spherical regions around each of the next-event detector points and tally the number of collisions within each such region; i.e., score the number of collisions occurring in the vicinity of each next-event detector.

   a. Using eqn 7.14P, is the flux result estimated with the detector consistent with the number of collisions calculated? Show how the results vary with the size of the spherical regions.

   b. Make the boundaries of your spherical regions into both standard and expectation surface-crossing detectors. How do these two flux estimates compare with the next-event estimator results? Has your calculation provided a thorough sampling of the important regions of phase space for each of the detectors?

Solution Part a

The idea of incorporating a bounding surface about a next-event estimator (NEE) in order to define a region of space near the detector point in which diagnostic scores could be collected is discussed in section 7.2 of the *Primer*. The benefit to be gained from employing such a surface, and using the resulting volume as a detector, is that it provides a convenient way to keep track of the number of collision events that occur close to the detector, or of particle tracks that pass close to the detector. The present exercise provides the opportunity to employ and evaluate the merit of such detectors in a simple problem for which, based on past calculations that include Examples 3.1, 5.1, and 7.1, we believe we know the answer to a high degree of accuracy.

In the first part of the exercise we place collision-density flux estimators in spherical regions surrounding the next-event detectors as used in Example 7.1 and compare the flux results from these volumetric detectors with the NEE results. Starting with the modified subroutines used in Example 7.1, as listed in Table 7.1P, it is relatively straightforward to produce such detectors. The list of subroutines modified for this calculation is given in Table 7.1.

Table 7.1. Subroutines Modified for Exercise 7.1a

Subroutine	Location
'Col'	Table 7.2
'Stats'	Table 7.3

The modified subroutine 'Col' is shown in Table 7.2. In this modified routine the changes indicated in Table 7.2P that are associated with the NEEs used in Example 7.1 are modified to reflect a re-positioning of the next-event detectors, as discussed below. In addition, changes are added to incorporate collision-density flux detectors about each of the NEEs. Thus the arrays 'cscore' and 'csumsq' are expanded in order to tally the pre-collision weights of the particles that are involved in collision events in each of the volumetric flux detectors, and the squares of these weights, respectively. The array 'ncolsn' is added to tally the number of collisions in each such volumetric detector. Tallies in the latter array help us determine whether we have achieved a reasonable sampling of the space near the various detectors. The scores in these volumetric detectors are recorded according to the zone number of the detector, which is assumed to be set by the geometry input as described below.

Subroutine 'Stats' as modified for this exercise is shown in Table 7.3. At entry point 'StatOne' the tally arrays for the collision-density detectors are set to zero and the volumes of the collision-density detectors are calculated. The radii of these detectors are obtained from the geometry input array 'sphr' (see Table 5.3P). The volume of the outermost collision density detector, which is centered about point detector eleven, at a radius of 9.9 mfp, requires special consideration, as discussed below. The coordinates of the NEEs are defined at entry point 'StatOne' by the arrays 'xdet,' 'ydet,' and 'zdet.' The changes under entry points 'StatLp' and 'StatELp' are similar to those of Table 7.3P. The final results, with estimates of standard deviations, are calculated at entry point 'StatEnd.' The changes under this entry point are also similar to those of Table 7.3P. Finally, the results are written to the output file.

Table 7.2. Subroutine 'Col' for Exercise 7.1a

```
SUBROUTINE COL                                                          !   1
REAL(8) FLTRN,delta,pi,dist2,tally
REAL(8) x,y,z,u,v,w,xo,yo,zo,uo,vo,wo,wate,age,energ                    !   3
COMMON/PART/x,y,z,u,v,w,xo,yo,zo,uo,vo,wo,wate,age,energ,nzcur,newzn,ngroup  !  4
REAL(8) dmfp,dtr,xsec,dcur                                              !   5
COMMON/TRACK/dmfp,dtr,xsec,dcur                                         !   6
REAL(8) sigt(20),c(20);  !  dimensions allow up to 20 different media       7
COMMON/GEOM/sigt,c; !sigt is total cross section, c is non-absorption prob   8
REAL(8) bscore(10),bsumsq(10),cscore(22),csumsq(22),bpart(10),cpart(22),&
 xdet(11),ydet(11),zdet(11),vol(11)
COMMON/STAT/bscore,bsumsq,cscore,csumsq,bpart,cpart,xdet,ydet,zdet,&
  vol,ncolsn(11)
pi=2.0d0*DACOS(0.0d0)  !  to obtain a double precision Pi
delta=dmfp/xsec                    !  distance traveled to collision         9
dtr=dcur+delta                     !  update total distance traveled        10
x=xo+u*dtr;  y=yo+v*dtr;  z=zo+w*dtr    !  update position                  11
!  collision-density scores in zones 2 through 12 (regions around dets 1-11)
IF(nzcur.NE.1) THEN
  cpart(nzcur+10)= cpart(nzcur+10)+wate
  ncolsn(nzcur-1) = ncolsn(nzcur-1)+1
ENDIF
wate=wate*c(nzcur)                 !  reduce wate by non-absorption probability  12
DO I=1,11                          !  NEE scores
  dist2=(x-xdet(i))**2+(y-ydet(i))**2+(z-zdet(i))**2  ! square of dist to detector
  tally=wate*EXP(-SQRT(dist2))/(4.*pi*dist2) ! score to detector
  cpart(i)=cpart(i)+tally
END DO
IF(wate.LT.0.1d0)THEN              !  if wate small play Russian roulette   13
    IF(wate.LT.FLTRN())THEN        !  if particle killed by RR              14
        nzcur=-1                   !  set nzcur=-1 to show particle killed  15
        RETURN                                                          !  16
    ENDIF                                                               !  17
    wate=1.0d0                     !  particle survived RR, increase wate   18
ENDIF                                                                   !  19
CALL ISOOUT                        !  assumes isotropic scatter in lab system  20
RETURN                                                                  !  21
END                                                                     !  22
```

By symmetry, the flux in the sphere of scattering material defined for this problem is a function of radius only. Thus to obtain an estimate of the flux at a given radius we can position a point detector anywhere on the surface of an imaginary sphere of this radius whose center is coincident with the point source. Because of this symmetry, in Example 7.1 we position all of the point detectors along the +Z axis. However, we are not constrained to placing the point detectors at these locations.

In the current exercise we wish to include collision-density detectors co-located with the NEEs by defining spherical regions about each of the point detectors. In so doing we may wish to ensure that these spherical regions do not interfere with one another; i.e., that they do not intersect. Each point in the geometry must be in one and only one zone, and if we wish to use geometric zones to define the collision-density spheres then the zones defined by such spheres must not overlap. On the other hand, it is possible to define scoring regions that do overlap. That is, we could define our scoring regions to be overlapping spheres provided such spherical regions are not defined as zones in the geometry. In this case collisions that occur in the overlapping regions would be scored in all of the detectors that are associated with such regions. For this exercise we use geometric zones to define the scoring regions and thus we require that the collision-density spheres do not overlap.

Table 7.3. Subroutine 'Stats' for Exercise 7.1a

```
SUBROUTINE Stats                                                          !    1
     REAL(8) bscore(10),bsumsq(10),cscore(22),csumsq(22),bpart(10),cpart(22),&
      xdet(11),ydet(11),zdet(11),vol(11)
     COMMON/STAT/bscore,bsumsq,cscore,csumsq,bpart,cpart,xdet,ydet,zdet,&
      vol,ncolsn(11)
     COMMON/IN/npart,nbatch                                               !    4
     REAL(8) x,y,z,u,v,w,xo,yo,zo,uo,vo,wo,wate,age,energ                 !    5
     COMMON/PART/x,y,z,u,v,w,xo,yo,zo,uo,vo,wo,wate,age,energ,nzcur,newzn,ngroup  ! 6
     REAL(8) tmp,var,stdev,pi,s
     REAL(8) sphcx(20),sphcy(20),sphcz(20),sphr(20),rpxmin(20),rpxmax(20),&
      rpymin(20),rpymax(20),rpzmin(20),rpzmax(20),bdin(40),bdout(40)
     COMMON/cg/sphcx,sphcy,sphcz,sphr,rpxmin,rpxmax,rpymin,rpymax,rpzmin,rpzmax,&
      bdin,bdout,nsph,nrpp,nbdy,nzones,nbz(40,41),nnext(40,40,40,2),bdtest,bdhit
ENTRY StatOne          ! entry point to initialize arrays for complete problem 8
     pi=2.0d0*DACOS(0.0d0)              ! to obtain a double precision Pi
     DO i = 1,11
       vol(i) = 4.*pi*sphr(i+1)**3/3.     !  vols of collision density detectors 1-11
     ENDDO
     IF(SPHR(12).gt.0.1)THEN      !   if rad > 0.1 recalculate vol for detector 11
       s=(sphr(1)**2-sphcy(12)**2-sphr(12)**2)/(2.0d0*sphcy(12))
       tmp=pi*(2.00d0*sphr(1)**3*(1.0d0-(sphcy(12)+s)/sphr(1))-&
        2.0d0*sphr(12)**3*(1.0d0-s/sphr(12))-sphcy(12)*(sphr(12)**2-s**2))/3.0d0
       vol(11) = vol(11)+tmp
     ENDIF
     DO I=1,11                       !   set NEE detector locations
       xdet(i)=sphcx(i+1); ydet(i)=sphcy(i+1);  zdet(i)=sphcz(i+1)
     ENDDO
     bscore=0.0d0;  bsumsq=0.0d0;  cscore=0.0d0;  csumsq=0.0d0
RETURN                                                                    !   10
ENTRY StatLp           ! entry point to initialize arrays for a particle      11
     bpart=0.0d0;  cpart=0.0d0                                            !   12
RETURN                                                                    !   13
ENTRY StatELp          ! entry point to store scores for a particle           14
     DO 80 i=1,22
       cscore(i)=cscore(i)+cpart(i)          !  store score
       csumsq(i)=csumsq(i)+cpart(i)**2       !  store square for variance calculation
  80 CONTINUE
RETURN                                                                    !   19
ENTRY StatEnd          ! entry point to calculate and print results           20
     tmp=DFLOAT(npart)                                                    !   21
     WRITE(16,12)
12   FORMAT(' NEE results by detector')
     DO 90 i=1,11
       var=csumsq(i)/tmp-(cscore(i)/tmp)**2              !  variance
       stdev=DSQRT(var)                                  !  standard deviation
       IF(i.lt.11)&
         WRITE(16,10)i,cscore(i)/tmp+DEXP(-DABS(zdet(i)))/(4.0d0*pi*zdet(i)**2),&
         cscore(i)/tmp,stdev/DSQRT(tmp)      !  add uncollided flux to first column
       IF(i.eq.11)&
         WRITE(16,10)i,cscore(i)/tmp+DEXP(-DABS(ydet(i)))/(4.0d0*pi*ydet(i)**2),&
         cscore(i)/tmp,stdev/DSQRT(tmp)      !  add uncollided flux to first column
90   CONTINUE
10   FORMAT(1x,i5,1p3D14.5)
     WRITE(16,11)
11   FORMAT(' collisions in zones 2-12 about dets 1-11')
```

Table 7.3. con't...

```
      DO 70 i = 1,11
        volfac = vol(I)
        var = csumsq(i+11)/tmp-(cscore(i+11)/tmp)**2
        stdev = DSQRT(var)
        WRITE(16,13)i,ncolsn(i),cscore(i+11)/tmp/volfac,&
        stdev/DSQRT(tmp)/volfac
13    FORMAT(1x,i5,i6,1p2d14.5)
 70   CONTINUE
   RETURN                                                    !     28
      END                                                    !     29
```

If we were to score all of our collision-density detectors in one calculation while retaining the positions of the NEEs along the +Z axis, we would have to restrict the size of the collision-density spheres in order to avoid interference. Because we do not wish to be constrained to collision-density regions with small radii, we must either perform a series of calculations in which we vary the radii of regions about individual detectors to avoid interference, or we must change the positions of the point detectors. In the latter case we may still have problems with interference between large-radius volumetric detectors positioned about point detectors one and two because of the proximity of these detectors to each other, but we have greater flexibility than that available while retaining the previous point detector locations.

For present purposes we place point detectors one through ten alternately along the positive- and negative-Z axis, thereby spacing them two mfp apart, except for detector one which is only 1.1 mfp from detector two and 1.9 mfp from detector three. With this configuration, except for detector one, we can use collision-density spheres with radii up to one mfp without such spheres overlapping. We locate detector eleven on the +Y axis and thus we are able to use a collision density sphere of large radius about this detector without risk of interfering with the other detectors. Detector one remains a special case. To avoid interference problems for the collision-density zones located around detectors one and two, the radius of the region about detector one is limited to 1.1 mfp minus the radius of the collision-density region about detector two. Care must also be taken to avoid interference between the spheres about detectors one and three. The detector positions chosen for this exercise are shown in Table 7.4.

We define spherical regions about the point detectors using the PFC geometry input, as described in Section 5.2.1 of the *Primer*. Starting with the geometry of Example 7.1, as given in Table 5.23P, we retain body one as the sphere of scattering material of radius $r_s = 10$ mfp. Bodies two through twelve are defined as small spheres placed about each point detector, beginning with detector one and ending with detector eleven. The detectors are numbered as shown in Table 7.4. We vary the radii of these collision-density spheres to determine the effect of such variation on the number of collisions that is recorded for each of the spherical volumes, and on the resulting flux estimates.

Table 7.4. Coordinates y and z of point detectors (x = 0 in all cases)

Detector #	y	z
1	0	-0.1
2	0	1
3	0	-2
4	0	3
5	0	-4
6	0	5
7	0	-6
8	0	7
9	0	-8
10	0	9
11	9.9	0

The geometry input file 'geom.txt' used for this problem is shown in Table 7.5. By restricting the radii of spheres two through eleven to values less than one mfp, we can define zones two through eleven as the interiors of these spheres. To place non-interfering spherical regions about detectors one and two we must restrict the radius of one sphere to a value less than 1.1 minus the radius of the other. Finally, since the radius of sphere twelve can be greater than 0.1, we must define the region about this sphere to be the intersection of sphere twelve with sphere one. Because the material in each of the collision-density detector zones is the same as that in the scattering sphere, the cross section input file 'xsects.txt' is the same as that in Table 5.24P except that the data are repeated twelve times, once for each of the non-void zones in the problem. As in Example 7.1, by defining the total cross section to be 1 cm^{-1} the 10-cm radius of the scattering sphere is equal to 10 mfp.

In order to use eqn 7.14P to obtain estimates of the flux in the collision-density detectors, as is required by the problem definition, we must define the volume of each of these detectors. The array 'vol' in the modified subroutine 'Stats' shown in Table 7.3 is used to store these volumes. The calculation of the volume of a collision-density detector around NEE eleven that involves a sphere centered at the detector and has a radius greater than 0.1 cm must account for the fact that this radius extends beyond the edge of the scattering sphere while the scattering material does not.

The geometry of detector eleven is shown in Figure 7.1. Here r_s is the radius of the large sphere of scattering material, r_{det} is the distance from the source to the detector point, and r_c is the radius of the collision-density detector sphere about point-detector eleven. We assume $r_{det} + r_c > r_s$. Let d be the distance between the point detector and the plane p defined by the intersection of the two spherical surfaces. This intersection is a circle of radius ρ, where

$$\rho^2 = r_s^2 - (r_{det} + d)^2 = r_c^2 - d^2$$

Table 7.5. Geometry input file 'geom.txt' for Exercise 7.1a assuming $r_c = 0.1$

0	There are no RPP bodies
12	There are 12 SPH bodies
0.0 0.0 0.0 10.0	Body 1 is the sphere of scattering material
0.0 0.0 -0.1 0.1	Body 2 is the volume detector about NEE 1
0.0 0.0 1.0 0.1	Body 3 is the volume detector about NEE 2
0.0 0.0 -2.0 0.1	Body 4 is the volume detector about NEE 3
0.0 0.0 3.0 0.1	Body 5 is the volume detector about NEE 4
0.0 0.0 -4.0 0.1	Body 6 is the volume detector about NEE 5
0.0 0.0 5.0 0.1	Body 7 is the volume detector about NEE 6
0.0 0.0 -6.0 0.1	Body 8 is the volume detector about NEE 7
0.0 0.0 7.0 0.1	Body 9 is the volume detector about NEE 8
0.0 0.0 -8.0 0.1	Body 10 is the volume detector about NEE 9
0.0 0.0 9.0 0.1	Body 11 is the volume detector about NEE 10
0.0 9.9 0.0 0.1	Body 12 is the volume detector about NEE 11
13	There are 13 zones in the geometry
12	Zone 1 includes 12 bodies
1 -2 -3 -4 -5 -6 -7 -8 -9 -10 -11 -12	Definition of zone 1
1	Zone 2 includes 1 body
2	Definition of zone 2 as body 2
— (entries for zones 3-11)	— insert analogous definitions for zones 3-11
2	Zone 12 includes two bodies
1 12	Zone 12 is the intersection of bodies 1 and 12
1	Zone 13 includes 1 body
-1	Definition of zone 13 as external void

Solving for d we obtain

$$d = \frac{r_s^2 - r_{det}^2 - r_c^2}{2r_{det}} \tag{7.1}$$

Recall that the volume of a cone with altitude h and base radius r is

$$V_{cone} = \frac{\pi}{3} r^2 h$$

The volume of a sector of a sphere of radius r, with the sector defined by the Z axis and the polar half-angle θ, is

$$V_{sph\,sec} = \frac{2\pi}{3} r^3 (1 - \cos\theta)$$

Thus, by inspection from Figure 7.1, we have for the volume of the portion of the collision-density sphere about NEE eleven that is inside the sphere of scattering material

$$V_{11} = V_{collsph} - V_{sphsec1} + V_{cone1} + V_{sphsec2} - V_{cone2}$$

Here $V_{collsph}$ is the total volume of the sphere of radius r_c and the other volume variables, with subscripts cone1, cone2, sphsec1, and sphsec2, refer to the volumes of the cones and spherical sectors defined by the angles θ_1 and θ_2 as shown in Figure 7.1. Using the above equations for the various volumes and eqn 7.1 for d, we find

$$V_{11} = \frac{4}{3}\pi r_c^3 + t$$

where

$$t = \frac{\pi}{3}\left\{2r_s^2\left[1 - \frac{r_{det} + d}{r_s}\right] - 2r_c^3\left[1 - \frac{d}{r_{det}}\right] - r_{det}\left(r_c^2 - d^2\right)\right\}$$

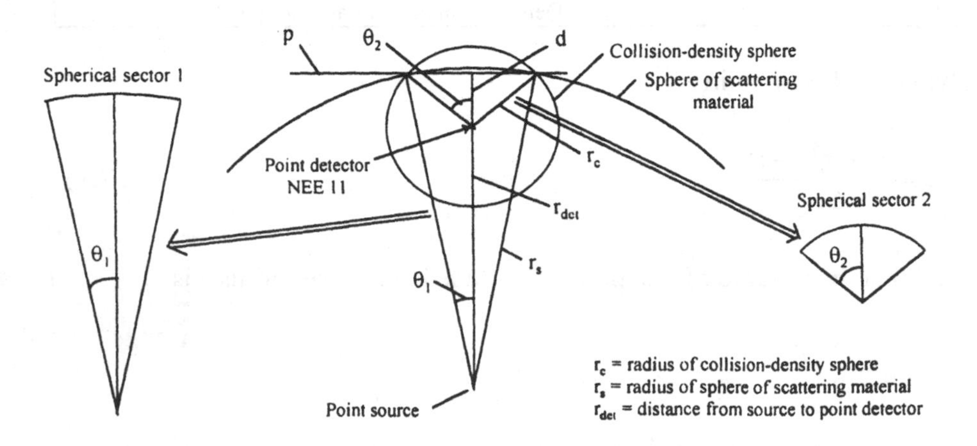

Figure 7.1. Diagram of Detector 11

The collision-density flux estimates obtained by running the code as described here for 10^5 source particles and for collision-density detector radii r_c of 1.0, 0.5, and 0.1 cm are shown in Figure 7.2 as a function of the distance between the source and the NEEs. This figure can be compared with Figures 3.3P and 7.2P. The NEE results obtained from this calculation are also shown. A separate calculation is made to obtain the collision-density result for detector one with $r_c = 1$ to avoid the zone overlap with detector two. These results are also shown in Table 7.6. The surface-crossing results from Table 7.16P are included for comparison.

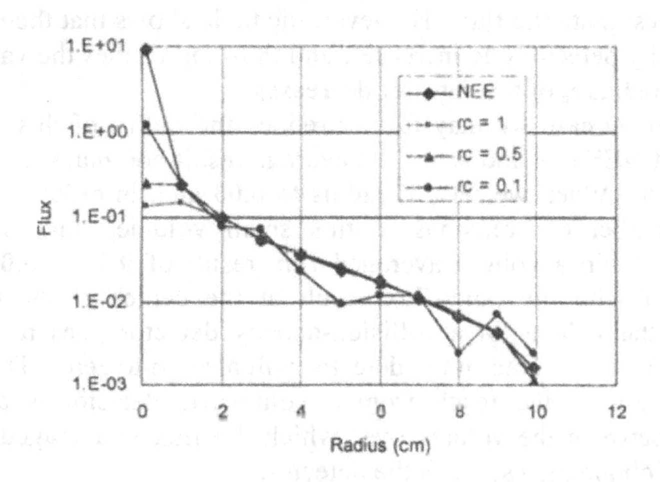

Figure 7.2. Flux vs Radius for Exercise 7.1a

Table 7.6. Results of Exercise 7.1a

det #	r_{det}	Comparisons: NEE results	Comparisons: Surface flux (Table 7.16P)	$r_c = 1$: # coll	$r_c = 1$: Collision-dens Flux	$r_c = 0.5$: # coll	$r_c = 0.5$: Collision-dens Flux	$r_c = 0.1$: # coll	$r_c = 0.1$: Collision-dens Flux
1	0.1	9.181±030	9.286±042	57502	0.1373±0011	13289	0.254±003	502	1.198±057
2	1	0.245±0091	0.241±002	63518	0.1516±0011	12748	0.243±003	103	0.246±027
3	2	0.0994±0046	0.0998±0005	41551	0.0992±0009	5175	0.0988±002	33	0.0788±0149
4	3	0.0587±0018	0.0575±0003	23760	0.0567±0007	2910	0.0556±0015	31	0.0740±0141
5	4	0.0360±0009	0.0371±0002	15722	0.0375±0006	1945	0.0371±0012	10	0.0239±0075
6	5	0.0240±0007	0.0255±0001	10995	0.0262±0005	1343	0.0256±0010	4	0.0095±0048
7	6	0.0158±0005	0.0174±82-4	7335	0.0175±0004	866	0.0165±0008	5	0.0119±0053
8	7	0.0110±0005	0.0118±55-4	4866	0.0116±0003	598	0.0114±0007	5	0.0119±0063
9	8	0.0068±0003	0.00759±38-4	3095	0.0074±0003	372	0.0071±0005	1	0.0024±0024
10	9	0.0043±0003	0.00422±22-4	1797	0.0043±0002	227	0.0043±0004	3	0.0072±0041
11	9.9	0.0017±0003	0.00164±93-5	561	0.0024±0002	59	0.0018±0003	1	0.0024±0024

From Table 7.6 we see that the collision-density results for detector one are smaller than the point-flux and surface-flux values for all of the r_c values shown. This is because the collision-density flux estimate is an average over the volume of the detector. In general we would not expect the flux averaged over a spherical volume to be equal to the flux at the center of the volume. This is particularly true for NEE one because the latter is located near the point source, where the flux is changing rapidly with radius. Even for a radius of $r_c = 0.1$ the collision-density result for detector one is low. It is obvious that a spherical, volumetric detector with a radius equal to the distance of its center from a point source is too large to obtain an accurate estimate of the flux at the center of the detector. To obtain results that are spatially accurate, the characteristic dimension of a volumetric detector should be significantly smaller than the spatial resolution to which one is attempting to estimate the flux. However, the table shows that the result for collision-density detector one increases, and thus approaches the value at its center, as the radius r_c of the detector decreases.

As an extreme case we may further reduce the radius of this spherical detector about NEE one and see if the average result continues to approach the point result. When we set this radius to 0.03 and, in order to obtain a reasonable number of collisions in this small volume, run 10^7 source particles, we obtain a volume-averaged flux result of 9.33 ± 0.09. This compares well with the point-flux result at the center of the detector. However, as the volume of a collision-density detector goes to zero the number of collisions scored in the detector will also go to zero. That is, the spatial resolution of the result from a volumetric detector is always a compromise between the volume over which the flux is averaged and the probability of obtaining a score in the detector.

The results obtained from the collision-density detectors about NEE two show the same trend as those for detector one in that the value for $r_c = 1$ is low compared with the point flux at the center of the detector. However, for $r_c \leq 0.5$ the collision-density result for detector two is in good agreement with the surface-crossing result. For detectors three through ten the results for both $r_c = 1.0$ and $r_c = 0.5$ are in good agreement with the previous results. For the distances from the source at which these detectors are located the mean value of the flux in a spherical region even as large as 1 mfp in radius is very close to the value at the center of the sphere. However, for the present calculation the number of collisions obtained in collision-density detectors three through ten for $r_c = 0.1$ is unacceptably small and the results should not be trusted.

The flux results for detector eleven vary unevenly as the radius of the collision-density detector decreases. Because of the distorted shape of this detector volume the result obtained for $r_c = 1$ is high. For detector radii greater than 0.1 mfp most of the volume of the detector is closer to the source than the NEE location, and the volume-averaged flux reflects this fact. The effect of the distortion in the detector volume for $r_c = 0.5$ is less than that for $r_c = 1$, and the result obtained for the former is much closer to the point-detector result than the latter. However, there are only 59 collisions in the detector for $r_c = 0.5$, so the result is not very accurate. Because only one collision is obtained in the detector for $r_c = 0.1$ the result for this radius has no merit.

According to eqn 7.14P (see also eqn 3.30P) the number of collisions n that will occur per unit time in a region of volume V with constant total cross section Σ_t can be estimated from the average flux ϕ in the region,

$$n = \phi \Sigma_t V \tag{7.2}$$

Using eqn 7.2 let us estimate the number of collisions that we would expect to obtain in the collision-density detectors used in this exercise based on our previous calculations. Using the surface-crossing flux results shown in Table 7.16P, as reproduced in Table 7.6, and assuming a source of 10^5 particles, we obtain the results shown in Table 7.7. The number of "expected" collisions is based on the assumption that the flux in each detector is constant and is equal to the value at the center of the detector. The difference between the predicted number of collisions and the actual number obtained in the present calculation varies in a manner consistent with the results of Table 7.6. Although no uncertainties are given in this table for the results from either eqn 7.2 or from the present calculation, one could assume the uncertainties are approximately equal to the square root of the number of collisions.

Table 7.7. Number of collisions expected in volume detectors 1 - 11 per 10^5 start particles based on boundary crossing fluxes from Table 7.16P, Exercise 7.1a

			Collisions in $r_c = 0.1$		Collisions in $r_c = 0.5$		Collisions in $r_c = 1.0$	
Det	radius (mfp)	Bdrx flux from Table 7.16P	Calc Eqn 7.2	Present calc	Calc Eqn 7.2	Present calc	Calc Eqn 7.2	Present calc
1	0.1	9.268±0.042	3880	502	485000	13289	3880000	57502
2	1.0	0.2406±0.0015	101	103	12600	12748	101000	63518
3	2.0	0.0998±0.00053	41.8	33	5230	5175	41800	41551
4	3.0	0.0575±0.00027	24.1	31	3010	2910	24100	23760
5	4.0	0.0371±0.00018	15.5	10	1940	1945	15500	15722
6	5.0	0.0255±0.00012	10.7	4	1340	1343	10700	10995
7	6.0	0.0174±8.2-5	7.29	5	911	866	7290	7335
8	7.0	0.0118±5.5-5	4.94	5	618	598	4940	4866
9	8.0	0.00759±3.8-5	3.18	1	397	372	3180	3095
10	9.0	0.00422±2.2-5	1.77	3	221	227	1770	1797
11	9.9	0.00164±9.3-6	0.687	1	55	59	381	561

Discussion

By examining the results of Table 7.7 along with the flux results shown in Table 7.6 we observe a close correlation between the number of collisions that occur in each volumetric detector and the flux estimates. When the number of collisions obtained in the calculation, as shown in Table 7.7, is greater than the number predicted by the boundary-crossing flux results in Table 7.6, the collision-density flux in Table 7.6 is high. When the number of collisions is lower than that predicted the flux result is low. Thus, for example, in detector ten with $r_c = 0.1$ we obtain three collisions while the predicted number is 1.77. The collision-density flux for this detector is estimated to be 0.0072 while the boundary-crossing flux is 0.00164. In detector nine for the same collision-density detector radius we obtain one collision versus the predicted number of 3.18. The boundary-crossing and collision-density flux estimates for this detector are 0.0024 and 0.00759, respectively. Finally, for detector six with $r_c = 0.5$, we obtain 1343 collisions versus 1340 predicted. The flux values are 0.0256 and 0.0255, respectively.

The best choice of radius for a spherical-volume, collision-density detector in the current problem would probably be $r_c \approx 0.1 r_{det}$ for a detector at distance r_{det} from a point source. We see from this problem that a collision-density detector of this size gives an accurate result for the flux at detector two. Intuitively we would expect this to remain true for real scattering and absorbing materials. In the current problem a volumetric detector of this size provides an adequate number of collisions to obtain a statistically meaningful answer in a reasonable amount of time. If the detector volume becomes small in proportion to the distance from the point source, as is the case for $r_c = 0.1$ at distances $r_{det} \geq 2$ mfp, accurate results will require either relatively long run times or the use of biasing techniques.

The rule of thumb mentioned on p. 36 of the *Primer*, that Monte Carlo results with an fsd greater than about 0.1 should be considered questionable, can be applied to the results of Table 7.6. Those results for which the fsd < 0.1 correspond to results for which the number of collisions is large. For example, in the last column of Table 7.6 ($r_c = 0.1$) the results for detectors three through eleven have fsd > 0.1 and are questionable both on this basis and on the basis of the number of collisions scored in the calculation. The result for detector two in this column (103 collisions, fsd ≈ 0.11) is marginal, while the result for detector one (502 collisions, fsd ≈ 0.048) is acceptable. The latter is true despite the fact that, consistent with the radius of this detector being large compared with the distance of its center from the point source, the result for detector one is low compared with that of the NEE located at its center.

Solution Part b

Boundary-crossing detectors are used in Examples 7.3 and 7.4. We use the techniques developed in those examples to solve the present exercise. We consider the two parts of the problem separately, first using the surfaces of the collision-density spheres introduced in part a of this exercise as standard surface-crossing detectors, and then using them as expectation surface-crossing detectors.

Standard Surface-Crossing Detectors

In Examples 7.3 and 7.4 the surface-crossing detectors consist of concentric spherical surfaces centered at the origin. As a result the determination of the normal to the boundary is straightforward. In the present problem the surface detectors are located at various positions within the sphere of scattering material. We must therefore determine the normal to each surface at the point the particle flight path passes through the surface based on the actual locations of the detectors.

If the center of a sphere of radius r_d is located at point $\mathbf{R}_d = (x_d,y_d,z_d)$, and a particle leaving the point **P** traveling in direction $\Omega = (u,v,w)$ intersects the sphere at the point $\mathbf{P}_1$ (see Figure 7.3, which is the general case of Figure 7.4P), the cosine of the angle between the normal to the sphere and the vector Ω at the point of intersection is

$$\mu = \Omega \bullet \mathbf{n} \tag{7.3}$$

where **n** is the unit outer normal on the sphere; i.e., **n** is a unit vector pointing in the same direction as a vector from the center of the sphere to the point of intersection of the particle track with the sphere. The point of intersection, (x,y,z), is given by eqns 3.44P and 3.49P, for s real and positive. Knowing this point of intersection we can find **n** by

$$\mathbf{n} = \frac{x - x_d}{r_d}\mathbf{i} + \frac{y - y_d}{r_d}\mathbf{j} + \frac{z - z_d}{r_d}\mathbf{k}$$

and thus by eqn 7.3 we know μ. If $\mu < 0$ the particle is entering the sphere and if $\mu > 0$ the particle is leaving the sphere.

The point of intersection of a flight path in PFC with a zone boundary is calculated in the library version of subroutine 'Bdrx' on line 15 (see Table 5.19P). The number of the zone in which the particle is traveling when it encounters the boundary is 'nzcur.' The zone number of the region being entered upon passing through the boundary just encountered is 'newzn.' Therefore the coding in subroutine 'Bdrx' provides the information needed to determine whether the boundary encountered is that of one of our spherical detectors, and if so which one, as well as to determine the coordinates of the point of the intersection of the flight path with the detector boundary. Equation 3.49P has already been solved and we need only use the results provided. Once we determine that a particle flight path intersects one of our spherical regions about a point detector we can easily score the boundary crossing in a detector using eqns 7.23P and 7.39P.

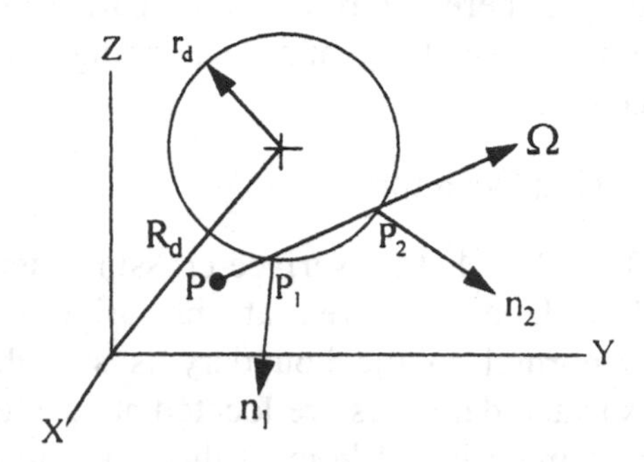

Figure 7.3. Flight path intersection with sphere

The surface-crossing detector about NEE eleven consists of portions of two spherical surfaces. The outer boundary of this detector is coincident with the outer radius of the scattering sphere when the radius of the detector is greater than 0.1 cm. This outer surface must be included separately in the boundary scoring for this detector. That is, to find the normal at the point of intersection of a flight path with the detector surface we need to determine which of the two spheres the path has encountered. When sphere 12 is encountered we find the normal as described above, and when the exterior of the scattering sphere is encountered we find the normal in the manner described in Example 7.3.

The list of subroutines modified to solve this first part of Exercise 7.1b is given in Table 7.8. The changes to 'Bdrx', shown in Table 7.9, are similar to those of Example 7.3, as shown in Table 7.14P. In this exercise we are again defining the collision-density spheres used in Exercise 7.1a, as shown in Table 7.5. Thus we assume zones two through twelve in this calculation correspond to the collision-density detectors one through eleven of that exercise. We continue this numbering by assuming the spherical surfaces of these collision density detectors correspond to standard surface-crossing detectors one through eleven. Thus the coding in Table 7.9 identifies particle flight paths that enter or exit zones two through twelve. When such a flight path has been identified the appropriate detector is scored using either eqn 7.23P or 7.39P. Obviously a single flight path can both enter and exit a zone. In this case, as always, both events must be scored. During the tracking of each particle the standard boundary scores are stored in the array 'bpart.' These contributions are summed for all particles at entry point 'StatElp' in subroutine 'Stats.'

The modified version of subroutine 'Col' is shown in Table 7.10. Changes have been made in this routine in order to accommodate the re-dimensioning of the 'bscore' and 'bsumsq' arrays in common 'stat.' The collision-density scores used in part a of the exercise have been retained so that we repeat the previous calculation as we run these new routines.

Table 7.8. Subroutines modified for Exercise 7.1b

Subroutine	Location
'Bdrx'	Table 7.9
'Col'	Table 7.10
'Stats'	Table 7.11

Table 7.9. Subroutine 'Bdrx' modified for Exercise 7.1b

```
      SUBROUTINE BDRX                                                        !   1
      REAL(8) delta,tmp,ctheta,xnorm,ynorm,znorm
      REAL(8) sphcx(20),sphcy(20),sphcz(20),sphr(20),rpxmin(20),rpxmax(20),&    !   3
       rpymin(20),rpymax(20),rpzmin(20),rpzmax(20),bdin(40),bdout(40)           !   4
      LOGICAL bdtest(40), bdhit(40)                                             !   5
      COMMON/cg/sphcx,sphcy,sphcz,sphr,rpxmin,rpxmax,rpymin,rpymax,rpzmin,rpzmax,& !  6
       bdin,bdout,nsph,nrpp,nbdy,nzones,nbz(40,41),nnext(40,40,40,2),bdtest,bdhit !  7
      REAL(8) x,y,z,u,v,w,xo,yo,zo,uo,vo,wo,wate,age,energ                     !   8
      COMMON/PART/x,y,z,u,v,w,xo,yo,zo,uo,vo,wo,wate,age,energ,nzcur,newzn,ngroup !  9
      REAL(8) dmfp,dtr,xsec,dcur                                                !  10
      COMMON/TRACK/dmfp,dtr,xsec,dcur                                           !  11
      REAL(8) bscore(22),bsumsq(22),cscore(22),csumsq(22),bpart(22),cpart(22),&
       xdet(11),ydet(11),zdet(11),vol(11)
      COMMON/STAT/bscore,bsumsq,cscore,csumsq,bpart,cpart,xdet,ydet,zdet,&
       vol,ncolsn(11),nbdry(22)
      delta=dtr-dcur                    ! delta--distance traveled to reach boundary  12
      dcur=dtr                          ! update current distance traveled            13
      dmfp=dmfp-delta*xsec              ! subtract current distance in mfp from dmfp  14
      x=x+u*delta;  y=y+v*delta;  z=z+w*delta     ! update position                   15
! find bdrx detector number - zone 1 is the scattering sphere, 2-12 are dets
      IF(newzn.LT.nzones)THEN        ! zone 13 = nzones is the external void
        idet = newzn-1
        IF(newzn.eq.1) idet = nzcur-1
        nbdry(idet) = nbdry(idet) + 1
        xnorm = (x-xdet(idet))/sphr(idet+1) ! normal to detector at intersection (x,y,z)
        ynorm = (y-ydet(idet))/sphr(idet+1)
        znorm = (z-zdet(idet))/sphr(idet+1)
        ctheta=xnorm*u+ynorm*v+znorm*w          ! cosine of angle with sphere normal
        IF(DABS(ctheta).LT.0.01d0)ctheta=0.005d0      ! correction for small angles
        bpart(idet)=bpart(idet)+wate/DABS(ctheta)     ! score boundary crossing
      ELSEIF(nzcur.EQ.12)THEN        ! score if bdry crossing in det 11 is leakage
        idet=11
        nbdry(idet) = nbdry(idet) + 1
        xnorm = (x-sphcx(1))/sphr(1)      ! normal to detector at intersection (x,y,z)
        ynorm = (y-sphcy(1))/sphr(1)
        znorm = (z-sphcz(1))/sphr(1)
        ctheta=xnorm*u+ynorm*v+znorm*w          ! cosine of angle with sphere normal
        IF(DABS(ctheta).LT.0.01d0)ctheta=0.005d0      ! correction for small angles
        bpart(idet)=bpart(idet)+wate/DABS(ctheta)     ! score boundary crossing
      ENDIF
      nzcur=newzn                       ! change identifier of current zone           16
      IF(nzcur.LT.nzones)RETURN         ! if not at outer zone then return            17
      nzcur=-1                          ! set nzcur=-1 if in outer zone (for escape)  18
      RETURN                                                                   !  19
      END                                                                      !  20
```

The modified subroutine 'Stats' is shown in Table 7.11. This routine is similar to that used in Example 7.3, as shown in Table 7.15P. The current exercise uses both the 'bscore' and 'cscore' arrays, along with their squares and particle tallies 'bsumsq', 'csumsq,' 'bpart,' and 'cpart.' It also uses event counters for these detectors stored in the array 'nbdry.' These arrays are zeroed at entry 'StatOne' or 'StatLp,' as appropriate.

Table 7.10. Modified Subroutine 'Col' for Exercise 7.1b

```
      SUBROUTINE COL                                                          !       1
      REAL(8) FLTRN,delta,pi,dist2,tally
      REAL(8) x,y,z,u,v,w,xo,yo,zo,uo,vo,wo,wate,age,energ                    !       3
      COMMON/PART/x,y,z,u,v,w,xo,yo,zo,uo,vo,wo,wate,age,energ,nzcur,newzn,ngroup   !  4
      REAL(8) dmfp,dtr,xsec,dcur                                              !       5
      COMMON/TRACK/dmfp,dtr,xsec,dcur                                         !       6
      REAL(8) sigt(20),c(20);  ! dimensions allow up to 20 different media            7
      COMMON/GEOM/sigt,c; !sigt is total cross section, c is non-absorption prob     8
      REAL(8) bscore(22),bsumsq(22),cscore(22),csumsq(22),bpart(22),cpart(22),&
       xdet(11),ydet(11),zdet(11),vol(11)
      COMMON/STAT/bscore,bsumsq,cscore,csumsq,bpart,cpart,xdet,ydet,zdet,&
        vol,ncolsn(11),nbdry(22)
      pi=2.0d0*DACOS(0.0d0)  !  to obtain a double precision Pi
      delta=dmfp/xsec                    !  distance traveled to collision            9
      dtr=dcur+delta                     !  update total distance traveled           10
      x=xo+u*dtr;  y=yo+v*dtr;  z=zo+w*dtr       !  update position                  11
 ! score collisions in zones 2 through 12 (regions around dets 1-11)
      IF(nzcur.NE.1) THEN
        cpart(nzcur+10)= cpart(nzcur+10)+wate
        ncolsn(nzcur-1) = ncolsn(nzcur-1)+1
      ENDIF
     wate=wate*c(nzcur)                 !  reduce wate by non-absorption probability  12
      DO i=1,11
        dist2=(x-xdet(i))**2+(y-ydet(i))**2+(z-zdet(i))**2  ! square of dist to detector
        tally=wate*EXP(-SQRT(dist2))/(4.*pi*dist2) ! score to detector
        cpart(i)=cpart(i)+tally
      END DO
      IF(wate.LT.0.1d0)THEN              !  if wate small play Russian roulette      13
          IF(wate.LT.FLTRN())THEN        !  if particle killed by RR                 14
              nzcur=-1                   !  set nzcur=-1 to show particle killed     15
              RETURN                                                          !      16
          ENDIF                                                               !      17
          wate=1.0d0                     !  particle survived RR, increase wate      18
      ENDIF                                                                   !      19
      CALL ISOOUT                        !  assumes isotropic scatter in lab system  20
      RETURN                                                                  !      21
      END                                                                     !      22
```

The collision-density and boundary-crossing tally arrays are summed at entry point 'StatELp' in subroutine 'Stats.' The mean values and standard deviations for the next-event, collision-density, and boundary-crossing tallies are calculated and recorded at entry point 'StatEnd.' The boundary tallies 'bscore' and 'bsumsq' are dimensioned 22 because this exercise calls for both standard and expectation boundary crossing scores. We anticipate using the subscript values 12-22 in these arrays after adding the expectation tallies for detectors 1-11. As stated above, we tally the boundary-crossing scores using the arrays 'bpart,' 'bscore,' and 'bsumsq.' The detector numbers used in Exercise 7.1a are used to designate the standard boundary crossing detectors as subscripts i = 1,11.

The area of surface-crossing detector eleven is not simply the surface of a sphere when the radius of the detector is greater than 0.1. To determine the surface area we consider the geometry shown in Figure 7.1. Again we assume $r_{det} + r_c > r_s$, and define d as the distance between the detector point and the plane defined by the intersection of the two spheres. Thus the surface area of detector eleven is $4\pi r_c^2$ minus the area of the zone of sphere twelve defined by $\theta = \theta_1$ plus the area of the zone of sphere one defined by $\theta = \theta_2$ (see Figure 7.1). The area of the zone of a sphere of radius r defined by polar angle θ is given by

$$A(\theta) = 2\pi r^2 (1 - \cos\theta)$$

Thus we have

$$A_{11} = 4\pi r_c^2 - 2\pi r_c^2 \left(1 - \frac{d}{r_c}\right) + 2\pi r_s^2 \left(1 - \frac{r_{det} + d}{r_s}\right)$$

where d is given by eqn 7.1.

The results obtained by running PFC with these modified routines, using 10^6 start particles and collision-density sphere radii of 0.1, 0.3, and 0.5 mfp except for detector one, are shown in Tables 7.12-14. As one may expect, there are many more boundary crossings than collisions scored in each detector, and the results for the surface-crossing detectors are generally good.

In part a of this exercise we conclude that a reasonable radius to choose for a collision density detector located a distance r_{det} from a source is $r_d \approx 0.1 r_{det}$. Using this relationship to set the radii of the collision-density and boundary-crossing detectors we obtain the boundary fluxes shown in Table 7.15 for 4×10^6 start particles. The number of particle tracks crossing each detector boundary, as well as the flux estimate and standard deviation, are shown. Except for detector eleven, these flux results compare well with those of Table 7.6 and Table 7.16P. For detector eleven, because of the marked asymmetry of the detector, the flux estimates from volume and surface-crossing estimators are not equal to the point flux at the location of NEE eleven. Furthermore, the average flux over the surface of the detector is different from the average flux over the volume of the detector.

Table 7.11. Subroutine 'Stats' for Exercise 7.1b

```
SUBROUTINE Stats                                                        !    1
     REAL(8) bscore(22),bsumsq(22),cscore(22),csumsq(22),bpart(22),cpart(22),&
      xdet(11),ydet(11),zdet(11),vol(11)
     COMMON/STAT/bscore,bsumsq,cscore,csumsq,bpart,cpart,xdet,ydet,zdet,&
       vol,ncolsn(11),nbdry(22)
     COMMON/IN/npart,nbatch                                             !    4
     REAL(8) x,y,z,u,v,w,xo,yo,zo,uo,vo,wo,wate,age,energ               !    5
     COMMON/PART/x,y,z,u,v,w,xo,yo,zo,uo,vo,wo,wate,age,energ,nzcur,newzn,ngroup  !  6
     REAL(8) tmp,var,stdev,pi,s
     REAL(8) sphcx(20),sphcy(20),sphcz(20),sphr(20),rpxmin(20),rpxmax(20),&
      rpymin(20),rpymax(20),rpzmin(20),rpzmax(20),bdin(40),bdout(40)
     COMMON/cg/sphcx,sphcy,sphcz,sphr,rpxmin,rpxmax,rpymin,rpymax,rpzmin,rpzmax,&
      bdin,bdout,nsph,nrpp,nbdy,nzones,nbz(40,41),nnext(40,40,40,2),bdtest,bdhit
ENTRY StatOne          ! entry point to initialize arrays for complete problem  8
     pi=2.0d0*DACOS(0.0d0)              ! to obtain a double precision Pi
     DO i = 1,11
       vol(i) = 4.*pi*sphr(i+1)**3/3.      ! vols of collision density detectors 1-11
     ENDDO
     IF(SPHR(12).gt.0.1)THEN     !  if rad > 0.1 recalculate vol for detector 11
       s=(sphr(1)**2-sphcy(12)**2-sphr(12)**2)/(2.0d0*sphcy(12))
       tmp=pi*(2.00d0*sphr(1)**3*(1.0d0-(sphcy(12)+s)/sphr(1))-&
        2.0d0*sphr(12)**3*(1.0d0-s/sphr(12))-sphcy(12)*(sphr(12)**2-s**2))/3.0d0
       vol(11) = vol(11)+tmp
     ENDIF
     DO i=1,11
       xdet(i)=sphcx(i+1); ydet(i)=sphcy(i+1);  zdet(i)=sphcz(i+1)
     ENDDO
     bscore=0.0d0; bsumsq=0.0d0; cscore=0.0d0; csumsq=0.0d0; nbdry = 0
RETURN                                                                  !   10
ENTRY StatLp              ! entry point to initialize arrays for a particle    11
     bpart=0.0d0; cpart=0.0d0                                           !   12
RETURN                                                                  !   13
ENTRY StatELp             ! entry point to store scores for a particle         14
     DO 80 i=1,22
       bscore(i)=bscore(i)+bpart(i)          ! store score
       bsumsq(i)=bsumsq(i)+bpart(i)**2       ! store square for variance calculation
       cscore(i)=cscore(i)+cpart(i)          ! store score
       csumsq(i)=csumsq(i)+cpart(i)**2       ! store square for variance calculation
  80 CONTINUE
RETURN                                                                  !   19
ENTRY StatEnd             ! entry point to calculate and print results         20
     tmp=DFLOAT(npart)                                                  !   21
     WRITE(16,12)
 12  FORMAT(' NEE results by detector')
     DO 90 i=1,11
       var=csumsq(i)/tmp-(cscore(i)/tmp)**2             ! variance
       stdev=DSQRT(var)                                 ! standard deviation
       IF(i.lt.11)&
         WRITE(16,10)i,cscore(i)/tmp+DEXP(-DABS(zdet(i)))/(4.0d0*pi*zdet(i)**2),&
         cscore(i)/tmp,stdev/DSQRT(tmp)      ! add uncollided flux to first column
       IF(i.eq.11)&
         WRITE(16,10)i,cscore(i)/tmp+DEXP(-DABS(ydet(i)))/(4.0d0*pi*ydet(i)**2),&
         cscore(i)/tmp,stdev/DSQRT(tmp)      ! add uncollided flux to first column
 90  CONTINUE
 10  FORMAT(1x,i5,1p3D14.5)
     WRITE(16,11)
 11  FORMAT(' collisions in zones 2-12 about dets 1-11')
 13  FORMAT(1x,i5,i6,1p2d14.5)
     DO 70 i = 1,11
       volfac = vol(I)
       var = csumsq(i+11)/tmp-(cscore(i+11)/tmp)**2
       stdev = DSQRT(var)
       WRITE(16,13)i,ncolsn(i),cscore(i+11)/tmp/volfac,&
       stdev/DSQRT(tmp)/volfac
```

Table 7.11. con't...

```
  70  CONTINUE
! bdrx detector results
      WRITE(16,15)
  15  FORMAT(' boundary crossing dets, no. of scores, flux, stdev')
      DO 170 i = 1,22
        var = bsumsq(i)/tmp-(bscore(i)/tmp)**2
        stdev = DSQRT(var)
        irad = i
        IF(i.gt.11)irad = i-11
        area = 4.*pi*sphr(irad+1)**2;   !  surface area of spherical detectors
        IF(irad.eq.11.and.sphr(12).gt.0.1)THEN   !  correction for detector 11
          s=(sphr(1)**2-sphcy(12)**2-sphr(12)**2)/(2.0d0*sphcy(12))
          area = area-2.*pi*sphr(12)**2*(1.-s/sphr(12))&
         +2.*pi*sphr(1)**2*(1-(sphcy(12)+s)/sphr(1))
        ENDIF
        WRITE(16,13)i,nbdry(i),bscore(i)/tmp/area,stdev/DSQRT(tmp)/area
 170  CONTINUE
RETURN                                                              !    28
      END                                                           !    29
```

Table 7.12. Results for Standard Surface Detector, Exercise 7.1b, 10^6 particles

			Collision Detector		Boundary Crossing Det	
det	NEE	rad	colls	results	# cross	results
1	9.22±.015	0.03	1031	9.12±2.9-1	53228	9.42±9.6-2
2	2.42-1±6-3	0.1	1006	2.40-1±8-3	15046	2.40-1±5-3
3	9.77-2±1-3	0.1	403	9.62-2±5-3	6056	9.58-2±3-3
4	5.93-2±1-3	0.1	277	6.61-2±4-3	3546	5.52-2±2-3
5	3.72-2±4-4	0.1	169	4.03-2±3-3	2498	3.87-2±1-3
6	2.56-2±7-4	0.1	100	2.39-2±3-3	1646	2.52-2±1-3
7	1.69-2±2-4	0.1	72	1.72-2±2-3	1088	1.62-2±9-4
8	1.30-2±2-3	0.1	52	1.24-2±2-3	696	9.95-3±6-4
9	7.11-3±2-4	0.1	20	4.77-3±1-3	428	7.32-3±7-4
10	4.52-3±3-4	0.1	23	5.49-3±1-3	264	4.23-3±5-4
11	1.61-3±6-5	0.1	9	2.15-3±7-4	120	1.89-3±3-4

Table 7.13. Results for Standard Surface Detector, Exercise 7.1b, 10^6 particles

			Collision Detector		Boundary Crossing Det	
det	NEE	rad	colls	results	# cross	results
1	9.22±.015	0.03	1031	9.12±2.9-1	53228	9.42±9.6-2
2	2.42-1±6-3	0.3	27210	2.41-1±1.8-3	136168	2.44-1±1.9-3
3	9.77-2±1-3	0.3	11075	9.79-2±1-3	55708	9.91-2±1-3
4	5.93-2±1-3	0.3	6537	5.78-2±9-4	32944	5.78-2±9-4
5	3.72-2±4-4	0.3	4311	3.81-2±7-4	21366	3.78-2±7-4
6	2.56-2±7-4	0.3	2838	2.51-2±6-4	14542	2.53-2±6-4
7	1.69-2±2-4	0.3	1965	1.74-2±5-4	9786	1.72-2±5-4
8	1.30-2±2-3	0.3	1295	1.15-2±4-4	6646	1.18-2±3-4
9	7.11-3±2-4	0.3	837	7.40-3±3-4	4228	7.47-3±3-4
10	4.52-3±3-4	0.3	516	4.56-3±3-4	2290	4.12-3±2-4
11	1.61-3±6-5	0.3	150	1.80-3±2-4	984	1.93-3±1-4

Table 7.14. Results for Standard Surface Detector, Exercise 7.1b, 10^6 particles

			Collision Detector		Boundary Crossing Det	
det	NEE	rad	colls	results	# cross	results
1	9.22±.015	0.03	1031	9.12±2.9-1	53228	9.42±9.6-2
2	2.42-1±6-3	0.5	126750	2.42-1±9.6-4	381888	2.45-1±1.0-3
3	9.77-2±1-3	0.5	51844	9.90-2±6-4	155022	9.87-2±7-4
4	5.93-2±1-3	0.5	30268	5.78-2±5-4	90876	5.78-2±5-4
5	3.72-2±4-4	0.5	19859	3.79-2±4-4	58534	3.71-2±5-4
6	2.56-2±7-4	0.5	13428	2.56-2±3-4	40340	2.59-2±4-4
7	1.69-2±2-4	0.5	8924	1.70-2±3-4	27136	1.76-2±4-4
8	1.30-2±2-3	0.5	6068	1.16-2±2-4	18610	1.22-2±3-4
9	7.11-3±2-4	0.5	3888	7.43-3±2-4	11854	7.55-3±2-4
10	4.52-3±3-4	0.5	2203	4.21-3±1-4	6462	4.09-3±1-4
11	1.61-3±6-5	0.5	638	1.91-3±1-4	2652	1.90-3±7-5

Table 7.15. Results for Standard Surface Detector, Exercise 7.1b, 4 x 10^6 particles

			Collision Detector		Boundary Crossing Det	
det	NEE	rad	colls	results	# cross	results
1	9.25±1.3-2	0.01	167	9.97±7.8-1	23366	9.31±1.5-1
2	2.41-1±3-3	0.1	3998	2.39-1±4-3	60200	2.41-1±2-3
3	9.81-2±5-4	0.2	13327	9.94-2±1-3	99396	9.98-2±8-4
4	5.73-2±4-4	0.3	25934	5.73-2±4-4	130546	5.80-2±5-4
5	3.82-2±1-3	0.4	40606	3.79-2±3-4	150830	3.74-2±3-4
6	2.53-2±3-4	0.5	53599	2.56-2±2-4	160390	2.55-2±2-4
7	1.73-2±2-4	0.6	63039	1.74-2±1-4	157832	1.73-2±1-4
8	1.18-2±4-4	0.7	67493	1.17-2±7-5	145028	1.18-2±8-5
9	7.43-3±9-5	0.8	64317	7.50-3±5-5	121678	7.55-3±6-5
10	4.37-3±1-4	0.9	51626	4.23-3±3-5	85754	4.26-3±4-5
11	1.57-3±3-5	0.99	23138	2.55-3±3-5	46854	2.37-3±3-5

Expectation Surface-Crossing Detectors

The versions of subroutines 'Bdrx,' 'Col,' and 'Stats' in the first part of Exercise 7.1b include coding to allow for the additional detectors needed for the expectation surface-crossing flux estimates. We now want to make use of the storage locations 12-22 in the arrays 'bscore,' 'bsumsq,' 'bpart,' and 'nbdry.' The list of subroutines modified for the present calculation is shown in Table 7.16.

Table 7.16. Subroutines Modified for Exercise 7.1b with Standard and Expectation Surface-Crossing Detectors

Subroutine	Location
'Bdrx'	Table 7.9
'Col'	Table 7.10
'Dist'	Table 7.17
'Stats'	Table 7.11

To implement the expectation boundary-crossing detectors we need to make changes in subroutine 'Dist,' as was done in Example 7.4. The modified subroutine 'Dist' is shown in Table 7.17. Subroutine 'Dist' determines the flight path of a particle in mfp and prepares for testing whether the track intersects the bodies in the geometry. In Table 7.19P this routine is modified to determine which, if any, of the bodies used to define surface detectors are encountered by this flight path. Here we use the same logic to determine whether the flight path encounters any of the surfaces we have placed around each of the point detectors, and to score the appropriate detectors.

The results obtained from running PFC as modified for this calculation for both the standard and expectation surface-crossing detectors using 10^5 start particles are shown in Tables 7.18 and 7.19. Table 7.18 shows the results for $r_d = 0.1$ mfp (except $r_1 = 0.005$ mfp), while Table 7.19 shows those for $r_d = 0.1 r_{det}$. These results may be compared with the NEE and collision-density detector results shown in Table 7.7, as well as the surface-crossing results of Example 7.3 shown in Table 7.16P. All of the results are in reasonable agreement.

It is apparent that, except for detector one, the expectation results include many more scores than the standard boundary-crossing results, and have a smaller variance than the latter. Except for the outermost detector, the expectation results with $r_d = 0.1 r_{det}$ appear to compare better with the surface-crossing results of Table 7.16P than the results with $r_d = 0.1$.

Despite the small variances obtained for the present results, the expectation surface-crossing flux estimates by themselves provide no information about the actual number of collisions or surface crossings that occurred in or near the detectors. We can compare the current expectation results with other results, including both the standard boundary-crossing results and the collision-density results, and see that in most cases we have in fact sampled the volumes about the detectors with reasonable thoroughness. However, in the absence of these comparisons, or of other supporting data, we could not say whether we have confidence in the expectation results.

In this exercise the important regions of phase space for determining the flux at each detector are those in the vicinity of the detector. The tallies of the collisions that occur in these regions along with the boundary crossings of the surfaces enclosing the regions provide a means of evaluating the adequacy of the sampling. In general, however, problems need to be evaluated on a case-by-case basis to determine the important regions of phase space. In shielding calculations, for example, streaming through shield penetrations may be very important to the result. The important regions of phase space for such calculations would be those that contribute to the streaming.

Table 7.17. Subroutine 'Dist' for Exercise 7.1b with Standard and Expectation Surface Crossing Detectors

```
      SUBROUTINE DIST                                                            !  1
      REAL(8) FLTRN,r,xi,yi,zi,ctheta
      REAL(8) dmfp,dtr,xsec,dcur                                                 !  3
      COMMON/TRACK/dmfp,dtr,xsec,dcur                                            !  4
      REAL(8) x,y,z,u,v,w,xo,yo,zo,uo,vo,wo,wate,age,energ                       !  5
      COMMON/PART/x,y,z,u,v,w,xo,yo,zo,uo,vo,wo,wate,age,energ,nzcur,newzn,ngroup !  6
      REAL(8) sphcx(20),sphcy(20),sphcz(20),sphr(20),rpxmin(20),rpxmax(20),&     !  7
       rpymin(20),rpymax(20),rpzmin(20),rpzmax(20),bdin(40),bdout(40)           !  8
      LOGICAL bdtest(40), bdhit(40)                                              !  9
      COMMON/cg/sphcx,sphcy,sphcz,sphr,rpxmin,rpxmax,rpymin,rpymax,rpzmin,rpzmax,& ! 10
       bdin,bdout,nsph,nrpp,nbdy,nzones,nbz(40,41),nnext(40,40,40,2),bdtest,bdhit ! 11
      REAL(8) bscore(22),bsumsq(22),cscore(22),csumsq(22),bpart(22),cpart(22),&
       xdet(11),ydet(11),zdet(11),vol(11)
      COMMON/STAT/bscore,bsumsq,cscore,csumsq,bpart,cpart,xdet,ydet,zdet,&
        vol,ncolsn(11),nbdry(22)
      REAL(8) sigt(20),c(20)   !  dimensions allow up to 20 different media
      COMMON/GEOM/sigt,c  !sigt is total cross section, c is non-absorption prob
      r=FLTRN()                        ! pick a random number                  12
      dmfp=-DLOG(r)                    ! distance traveled in mean free paths  13
      xo=x;  yo=y;  zo=z               ! store "starting" position for track   14
      dcur=0.0D0                       ! initialize distance currently traveled 15
      DO 50 i=1,nbdy                                                             ! 16
        bdtest(i)=.FALSE.              ! beginning a new track, no bodies yet tested 17
   50 CONTINUE                                                                   ! 18
      bdtest(1)=.TRUE.            !  score expectation boundary crossing detectors
      CALL SPH(1)
      DO 100 i = 2,12             !  detector numbers 12-22 correspond to bodies 2-12
        bdtest(i) = .true.
        CALL SPH(i)
        IF(bdhit(i)) THEN
         idet = i+10
        nbdry(idet) = nbdry(idet)+1
          IF(i.NE.12.OR.(i.EQ.12.AND.bdout(12).LT.bdout(1))) THEN
            xi = x+bdout(i)*u
            yi = y+bdout(i)*v
            zi = z+bdout(i)*w
            xnorm = (xi-xdet(i-1))/sphr(i)
            ynorm = (yi-ydet(i-1))/sphr(i)
            znorm = (zi-zdet(i-1))/sphr(i)
            ctheta=(xnorm*u+ynorm*v+znorm*w)               ! cosine of angle with normal
            IF(DABS(ctheta).LT.0.01d0)ctheta=0.005d0       ! correction for small angles
            bpart(idet)=bpart(idet)+wate*EXP(-sigt(nzcur)*bdout(i))/DABS(ctheta)  ! score
          ELSE
            xi = x+bdout(1)*u
            yi = y+bdout(1)*v
            zi = z+bdout(1)*w
            xnorm = (xi-sphcx(1))/sphr(i)
            ynorm = (yi-sphcy(1))/sphr(i)
            znorm = (zi-sphcz(1))/sphr(i)
            ctheta=(xnorm*u+ynorm*v+znorm*w)               ! cosine of angle with normal
            IF(DABS(ctheta).LT.0.01d0)ctheta=0.005d0       ! correction for small angles
            bpart(idet)=bpart(idet)+wate*EXP(-sigt(nzcur)*bdout(1))/DABS(ctheta)  ! score
          ENDIF
          IF(bdin(i).gt.0.) THEN
          IF(i.NE.12.OR.(i.EQ.12.AND.bdin(12).LT.bdout(1))) THEN
            nbdry(idet) = nbdry(idet)+1
              xi=x+bdin(i)*u;  yi=y+bdin(i)*v;  zi=z+bdin(i)*w !  exit intersect point
              xnorm = (xi-xdet(i-1))/sphr(i)
              ynorm = (yi-ydet(i-1))/sphr(i)
              znorm = (zi-zdet(i-1))/sphr(i)
              ctheta=(xnorm*u+ynorm*v+znorm*w)             ! cosine of angle with normal
              IF(DABS(ctheta).LT.0.01d0)ctheta=0.005d0     ! correction for small angles
```

Table 7.17. cont...

```
              bpart(idet)=bpart(idet)+wate*EXP(-sigt(nzcur)*bdin(i))/DABS(ctheta) !score
            ENDIF
          ENDIF
        ENDIF
100   CONTINUE
      RETURN                                                                  ! 19
      END                                                                     ! 20
```

Table 7.18. Results of Standard and Expectation Surface-Crossing Detectors for r_d = 0.1 (except r_1 = 0.005), 10^5 particles

	Standard bdrx			Expectation bdrx		
det #	# events	Flux	sigma	# events	Flux	sigma
1	134	8.945	1.40	160	9.018	1.26
2	1436	2.309-1	1.73-2	4329	2.255-1	1.32-2
3	598	9.044-2	6.62-3	2729	9.262-2	7.51-3
4	396	6.194-2	5.73-3	2027	6.513-2	4.64-3
5	218	2.945-2	3.61-3	1506	3.477-2	2.94-3
6	160	2.308-2	2.99-3	1214	2.570-2	2.29-3
7	94	1.421-2	2.36-3	881	1.370-2	1.65-3
8	44	5.279-3	1.16-3	725	7.023-3	9.55-4
9	46	6.901-3	1.67-3	579	9.138-3	2.44-3
10	32	4.599-3	1.37-3	457	3.678-3	6.92-4
11	14	2.237-3	1.18-3	387	1.738-3	6.84-4

Table 7.19. Results of Standard and Expectation Surface-Crossing Detectors for $r_d = 0.1r_{det}$

	Standard Bdrx			Expectation Bdrx		
Det #	# events	Flux	sigma	# events	Flux	sigma
1	578	9.512	8.44-1	661	9.220	7.72-1
2	1436	2.309-1	1.73-2	4329	2.255-1	1.32-2
3	2496	9.492-2	3.83-3	10969	9.332-2	2.63-3
4	3230	5.642-2	4.07-3	18002	5.753-2	3.72-3
5	3694	3.839-2	2.39-3	24761	3.701-2	1.06-3
6	4086	2.652-2	9.30-4	30623	2.567-2	7.02-4
7	3882	1.703-2	1.04-3	34432	1.668-2	5.86-4
8	3520	1.097-2	4.14-4	37050	1.139-2	3.64-4
9	3054	8.067-3	6.01-4	38167	7.902-3	5.37-4
10	2172	4.111-3	1.96-4	37206	3.966-3	1.56-4
11	1218	2.591-3	2.02-4	34345	1.939-3	9.80-5

Exercise 7.2

Statement of the problem

2. Consider a spherical shell of material (see Figure 7.4.) Assume the radial thickness of the material is 10% of its inner radius. Assume further that the spherical shell is contained in a box whose inner surfaces are tangent to the outside of the shell and whose walls are of the same thickness as the shell.

Consider a monoenergetic problem with isotropic scatter in the laboratory system for which there is an isotropic point source of neutrons at the center of the shell. Assume a vacuum exists inside the shell and between the shell and the box. Determine the flux as a function of position on the exterior faces of the box under the following conditions:

a. The shell and box are composed of a pure absorber with a mean free path equal to the thickness of the shell.

b. The shell and box are composed of a pure scatterer with a mean free path equal to the thickness of the shell.

c. The shell is composed of a scattering and absorbing material with a mean free path equal to the thickness of the shell and a non-absorption probability of 0.5, and the box is composed of a scattering and absorbing material with a mean free path equal to half the thickness of the shell and a non-absorption probability of 0.9.

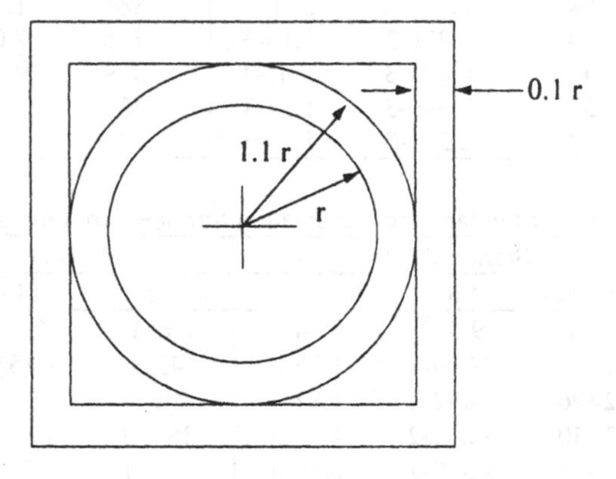

Figure 7.4. Geometry for Exercise 2

Solution

This problem represents a radiation source in a spherical-shell shield that is in turn contained in a hollow-cube shield. Because it adds some diagnostic value and is easy to implement, in addition to scoring the particle flux as a function of position on the surface of the cube we also score the particle current.

The three parts of Exercise 7.2 use the same geometry. Thus we can model the geometry and define a set of spatially dependent detectors that can be used with different cross sections to solve the various parts of the problem. By symmetry, all six faces of the box are identical and thus we can sum the leakage through all faces into a single tally to obtain the desired results. When we do this we must divide the final result by six to obtain the correct normalization. Further, within each face there is both horizontal, vertical, and 45-degree-diagonal lines of symmetry. Therefore we could use one eighth of a face to solve the entire problem. However, we instead use one quarter of a face, divided into a rectangular grid, to score the spatially dependent leakage current and flux. When we do this we add an additional factor to the normalization. When we sum over all faces we have the factor of six mentioned above. If we then sum over the four quarters of this face (summing the leakage from six faces onto one quarter of one face) we must divide our results by $6 \times 4 = 24$ to obtain the correct normalization.

We define a coordinate system on one face of the cube, with coordinate axes (η,ζ) parallel to one pair of adjacent edges, and with the origin of the coordinates at the center of the face. We then divide the positive quarter of the face, as defined by these coordinates, into N equal intervals along each of the axes such that the interval (n,m) defines the surface area of detector (n,m), as shown in Figure 7.5. We score the particle leakage and surface-crossing flux in each of these $N \times N$ bins.

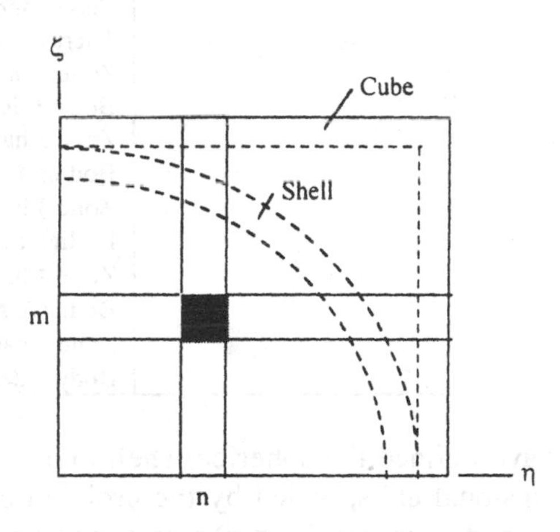

Figure 7.5. Detector Location on Face of Cube

Because the problem geometry is defined in terms of mfp based on the thickness of the spherical shell and box, we can make the actual dimensions of the geometry whatever we wish, consistent with Figure 7.4. Although the magnitude of the flux on the surface of the box will vary with the dimensions used, the shape of the flux will be the same independent of these dimensions. We therefore arbitrarily choose the inner radius of the spherical shell, r, to be ten units of length. Thus the shell and box each have thicknesses of one unit and the length of each edge of the box is 24 units.

We use two 'RPP' bodies and two 'SPH' bodies in the PFC file 'geom.txt' to define our geometry. The body numbers are determined by the order in which they are entered in the geometry input, with the 'RPP's preceding the 'SPH's. Starting from the inside out, we define five zones in this problem: 1) the interior of the sphere, 2) the spherical shell, 3) the region between the shell and the inside of the cube, 4) the walls of the cube, and 5) the exterior of the cube. The resulting geometry input for this problem is shown in Table 7.20. To avoid the possibility of roundoff errors creating confusion at the points at which the sphere is tangent to the inside of the cube, we define the inside of the cube to be slightly larger than the exterior of the spherical shell.

Table 7.20. Geometry Description for Exercise 7.2

2	There are two RPPs
-11.0001 11.0001 -11.0001 11.0001 -11.0001 11.0001	Location of faces of RPP 1
-12. 12. -12. 12. -12. 12.	Location of faces of RPP 2
2	There are two SPHs
0. 0. 0. 10.	Center and radius of SPH 1
0. 0. 0. 11.	Center and radius of SPH 2
5	There are five zones
1	Zone 1 has one body
3	Body 3 defines zone 1
2	Zone 2 has two bodies
4 -3	Bodies 4 and 3 define zone 2
2	Zone 3 has two bodies
1 -4	Bodies 1 and 4 define zone 3
2	Zone 4 has two bodies
2 -1	Bodies 2 and 1 define zone 4
1	Zone 5 has one body
-2	Body 2 defines zone 5

Because we have defined the spherical shell and cube thicknesses to be equal to 1.0 dimensional units, which by the problem definition must equal either one (parts a and b) or two (part c) mfp for our particles, according to eqn 3.5P we must set the total cross section either to 1.0 (parts a and b) or 2.0 (part c). The total cross sections are defined in the input file 'xsects.txt.' Table 7.21 shows the contents of these files for the three parts of this problem. Zones one and three contain void ($\Sigma_t = 0$) while zones two and four contain either pure absorbers (c = 0), pure scatterers (c = 1), or mixed absorption and scattering, all with a total cross section of 1.0 per unit length. The value of c for the void-filled zones is immaterial and has been set to zero.

Table 7.21. Cross Section File 'xsects.txt' for Exercise 7.2

Part a	Part b	Part c
0. 0.	0. 0.	0. 0.
0. 1.	1. 1.	0.5 1.
0. 0.	0. 0.	0. 0.
0. 1.	1. 1.	0.9 2.

To score the particles leaking from the geometry we must modify PFC. The routines modified for this calculation are listed in Table 7.22. These subroutine changes apply to all parts of Exercise 7.2 because only the cross sections have to be modified to solve the three parts of the problem.

Table 7.22. Modified subroutines for Exercise 7.2

Subroutine	Location
'Bdrx'	Table 7.23
'Stats'	Table 7.24

To score the particle current as a function of position in the quarter-face shown in Figure 7.5 we use the array 'bscore.' We doubly dimension both this array and the array used to score the square of the particle current for variance estimation, 'bsumsq.' We arbitrarily choose N = 10 for our spatially dependent scoring, and thus dimension these arrays 10 × 10. We score the surface-crossing flux in the array 'cscore' and similarly dimension that array and its associated square, 'csumsq.'

As stated, we wish to score all leaking particles in the same spatially dependent array, independent of the face, and independent of the quadrant of the face, through which they escape. Because 'Bdrx' already contains a test to determine whether a particular boundary crossing results in the escape of a particle, this subroutine is a logical place to do this. The modified subroutine 'Bdrx' is listed in Table 7.23. In 'Bdrx' the current zone 'nzcur' is set to -1 if the particle being tracked has escaped. At this point in the modified routine we insert a series of tests to find the face through which a particle has escaped. Since the outer edges of the box are located at ±12 dimensional units in Cartesian coordinates, we check to see which coordinate of the escaping particle has an absolute value approximately equal to this dimension. The particle will have escaped through one of the faces of the box that is perpendicular to this coordinate.

Knowing the face through which the particle has escaped we can then find the values of the indexes (n,m) corresponding to the coordinates of the (η,ζ) point at which the particle has escaped from the geometry. We do this by dividing the corresponding coordinate of the leaking particle by a quantity 1/N times the half-width of the face. However, to avoid possible roundoff errors, we have divided the coordinates by a quantity $(1/N + \delta)$, where δ is a small number. This ensures that the indexes cannot take on a value greater than N, where here N = 10. We then set the result for this coordinate to an integer. Because of truncation the integer plus one is the value of the bin index along that axis. Thus bin (1,1) is adjacent to the center of the face and the origin of the (η,ζ) coordinates, while bin (10,10) is adjacent to the corner of the box.

Once the indexes for the leaking particle have been determined it is simple to tally the current and flux estimates, along with their squares. The 'bscore' array simply sums the weights of the leaking particles and thus tallies the net current through the surface bins. The 'cscore' array uses either eqn 7.23P or 7.39P (the latter with $\varepsilon = 0.01$) to score the flux on the outer surface of the box. Finally the total number of particles escaping in each bin is tallied in the newly defined array 'ncount.' The modifications in 'Bdrx' have been coded for the specific dimensions selected for the geometry used here and must be changed if these dimensions are changed.

Subroutine 'Stats' has been modified to accommodate the revised dimensions of the scoring arrays and to calculate and normalize the current and flux estimates, as shown in Table 7.24. The normalization of the scores takes into account the fact that we are using only one fourth of one face to score all leaking particles, as explained above. Hence the scores 'bscore' and 'cscore' are normalized by dividing them by 24. The squares of these scores, 'bsumsq' and 'csumsq,' on the other hand, must be divided by 24^2 in order to maintain the correct normalization. After normalizing the leaking currents and surface fluxes the standard deviations are calculated and the results are stored on the output file. The total-leakage array 'ncount' is also written on the output file. Because no scores are recorded for collisions or other events inside the geometry, each particle can make only one score and the arrays 'bpart' and 'cpart' are not used. Therefore the tallies at entry point 'StatELp' used in the library version of 'Stats' have been deleted.

One simple method for checking the detector constructions, verifying the problem normalization, and validating the geometry is to run the calculation with a void in all of the zones. In such a calculation the source particles merely stream to the exterior of the geometry where they are tallied in the three arrays 'bscore,' 'cscore,' and 'ncount.' A free-streaming calculation can easily be done by setting all of the total cross sections to zero. The results from such a calculation can verify that the number of particles escaping is equal to the number of source particles. Therefore they show that all source particles successfully stream through the geometry and escape without becoming lost or otherwise encountering difficulties. Additionally, the diagonal symmetry of the problem means that the free-streaming flux in bin (n,m) should equal the flux in bin (m,n).

Table 7.23. Subroutine 'Bdrx' Modified for Exercise 7.2

```
      SUBROUTINE BDRX                                                       !   1
      REAL(8) delta                                                         !   2
      REAL(8) sphcx(20),sphcy(20),sphcz(20),sphr(20),rpxmin(20),rpxmax(20),& !   3
       rpymin(20),rpymax(20),rpzmin(20),rpzmax(20),bdin(40),bdout(40)       !   4
      LOGICAL bdtest(40), bdhit(40)                                         !   5
      COMMON/cg/sphcx,sphcy,sphcz,sphr,rpxmin,rpxmax,rpymin,rpymax,rpzmin,rpzmax,&  !  6
       bdin,bdout,nsph,nrpp,nbdy,nzones,nbz(40,41),nnext(40,40,40,2),bdtest,bdhit   !  7
      REAL(8) x,y,z,u,v,w,xo,yo,zo,uo,vo,wo,wate,age,energ                   !   8
      COMMON/PART/x,y,z,u,v,w,xo,yo,zo,uo,vo,wo,wate,age,energ,nzcur,newzn,ngroup   !  9
      REAL(8) dmfp,dtr,xsec,dcur                                            !  10
      COMMON/TRACK/dmfp,dtr,xsec,dcur                                       !  11
      REAL(8) bscore(10,10),bsumsq(10,10),cscore(10,10),csumsq(10,10)
      COMMON/STAT/bscore,bsumsq,cscore,csumsq,ncount(10,10)
      REAL(8) eta,zeta,xmu
      delta=dtr-dcur                  ! delta--distance traveled to reach boundary  12
      dcur=dtr                        ! update current distance traveled           13
      dmfp=dmfp-delta*xsec            ! subtract current distance in mfp from dmfp 14
      x=x+u*delta;  y=y+v*delta;  z=z+w*delta     ! update position                15
      nzcur=newzn                     ! change identifier of current zone          16
      if(nzcur.LT.nzones)RETURN       ! if not at outer zone then return           17
      nzcur=-1                        ! set nzcur=-1 if in outer zone (for escape) 18
      IF(DABS(x).gt.11.999d0) leak = 1     ! Find which face particle is leaking from
      IF(DABS(y).gt.11.999d0) leak = 2
      IF(DABS(z).gt.11.999d0) leak = 3
      Check_Leakage: SELECT CASE(leak)
        CASE(1)                   ! X-face
          eta =  DABS(y)
          zeta = DABS(z)
          xmu =  DABS(u)
        CASE(2)                   ! Y face
          eta =  DABS(x)
          zeta = DABS(z)
          xmu =  DABS(v)
        CASE(3)                   ! Z face
          eta =  DABS(y)
          zeta = DABS(x)
          xmu =  DABS(w)
      END SELECT Check_Leakage
      n = INT(eta/1.2001d0) + 1  !   use(eta,zeta) coordinates to score leak
      m = INT(zeta/1.2001d0) + 1
      IF(xmu.lt.0.01d0) xmu = 0.005d0
      bscore(n,m) = bscore(n,m) + wate
      bsumsq(n,m) = bsumsq(n,m) + wate**2
      cscore(n,m) = cscore(n,m) + wate/xmu
      csumsq(n,m) = csumsq(n,m) + (wate/xmu)**2
      ncount(n,m) = ncount(n,m) + 1
      RETURN                                                                !  19
      END                                                                   !  20
```

A selection of results obtained by setting the total cross sections to zero and running 10^6 source particles is shown in Table 7.25. We see from these results that, within statistics, the flux bins along the (η,ζ) axes of the problem exhibit the symmetry mentioned above. An examination of the complete PFC output file shows that this holds true for other values of n and m as well.

Table 7.24. Subroutine 'Stats' Modified for Exercise 7.2

```
SUBROUTINE Stats                                                              !  1
     REAL(8) bscore(10,10),bsumsq(10,10),cscore(10,10),csumsq(10,10)
     COMMON/STAT/bscore,bsumsq,cscore,csumsq,ncount(10,10)
     COMMON/IN/npart,nbatch                                                   !  4
     REAL(8) x,y,z,u,v,w,xo,yo,zo,uo,vo,wo,wate,age,energ                     !  5
     COMMON/PART/x,y,z,u,v,w,xo,yo,zo,uo,vo,wo,wate,age,energ,nzcur,newzn,ngroup  ! 6
     REAL(8) tmp,sum,var,stdev
ENTRY StatOne              ! entry point to initialize arrays for complete problem 8
     bscore=0.0d0;  bsumsq=0.0d0;  cscore=0.0d0;  csumsq=0.0d0                !  9
    ncount = 0  ; sum = 0.0d0
RETURN                                                                        ! 10
ENTRY StatLp               ! entry point to initialize arrays for a particle    11
RETURN                                                                        ! 13
ENTRY StatELp              ! entry point to store scores for a particle         14
RETURN                                                                        ! 19
ENTRY StatEnd              ! entry point to calculate and print results        20
     tmp=DFLOAT(npart)                                                        ! 21
     WRITE(16,*)' Leaking current'
     DO 1 i = 1,10
     DO 1 j = 1,10
       sum = sum + bscore(i,j)    !  total leakage
       bscore(i,j) = bscore(i,j)/1.2d0/1.2d0/24.d0   ! normalize to unit area
       bsumsq(i,j) = bsumsq(i,j)/1.2d0/1.2d0/24.d0/24.0d0
       var=bsumsq(i,j)/tmp-(bscore(i,j)/tmp)**2 !  variance of bscore distribution
       stdev=DSQRT(abs(var))                    !  standard deviation of bscore distr
       WRITE(16,2)i,j,ncount(i,j), bscore(i,j)/tmp,stdev/DSQRT(tmp) ! result & std dev
1    CONTINUE
     WRITE(16,*)'total leakage ',sum
     WRITE(16,*)' Surface flux'
     DO 3 i = 1,10
     DO 3 j = 1,10
       cscore(i,j) = cscore(i,j)/1.2d0/1.2d0/24.d0   ! normalize cscore
       csumsq(i,j) = csumsq(i,j)/1.2d0/1.2d0/24.d0/24.0d0
       var = csumsq(i,j)/tmp-(cscore(i,j)/tmp)**2
       stdev = DSQRT(DABS(var))
       WRITE(16,2)i,j,ncount(i,j),cscore(i,j)/tmp,stdev/DSQRT(tmp)
3    CONTINUE
2    FORMAT(2i5,i8,2x, 1p2e12.4)
     RETURN                                                                   ! 28
     END                                                                      ! 29
```

We can easily make some analytical estimates to compare with the PFC results for this simple case of free-streaming particles escaping from the problem geometry. For example, the flux at the center of each face must equal $\phi_o = (4\pi L^2)^{-1}$, where L = 12 is the distance from the source to the face of the box. Because we did not attempt to estimate this quantity in our PFC calculation, the closest we can come to this result is the flux estimate for bin (1,1). The center of this bin is at the coordinates (0.6,0.6) in the (η,ζ) coordinate system. Thus the distance from the source point to the center of this bin is about 12.03 units. Substituting this for L gives 5.499×10^{-4}. The PFC result, shown in Table 7.25, is $(5.522 \pm 0.048) \times 10^{-4}$. These results agree well within statistics, which are excellent with an fsd of about 0.009. As another example let us consider the flux in bin (10,10). The distance from the source point to the center of this bin is about 20.0978 units. Thus we estimate the free-streaming flux in this bin to be approximately 1.9701 ×

10^{-4}. The PFC result, not shown in Table 7.25, is $(1.959 \pm 0.037) \times 10^{-4}$ which once again, within statistics, is in agreement with the analytic result. The fsd for this PFC result, about 0.019, is larger than that of bin (1,1). This is a consequence of the fact that the solid angle of bin (10,10) is smaller, as viewed from the source point, than that of bin (1,1), and therefore should receive fewer scores in this streaming calculation than the latter.

Table 7.25. Selected Flux Results from Exercise 7.2 with $\Sigma_t = 0$ in All Zones

n	m	# parts	Flux	sigma
1	1	19022	5.5224-4	4.7731-6
1	2	18342	5.3774-4	4.7342-6
1	3	17449	5.2135-4	4.7075-6
1	4	16128	4.9521-4	4.6532-6
1	5	14426	4.5834-4	4.5565-6
1	6	12740	4.2115-4	4.4579-6
1	7	11282	3.8967-4	4.3854-6
1	8	9827	3.5565-4	4.2910-6
1	9	8525	3.2392-4	4.1978-6
1	10	7349	2.9338-4	4.0967-6
1	1	19022	5.5224-4	4.7731-6
2	1	18166	5.3257-4	4.7117-6
3	1	17561	5.2471-4	4.7225-6
4	1	16051	4.9288-4	4.6425-6
5	1	14583	4.6332-4	4.5809-6
6	1	12654	4.1851-4	4.4451-6
7	1	11138	3.8469-4	4.3576-6
8	1	9872	3.5726-4	4.3005-6
9	1	8542	3.2454-4	4.2017-6
10	1	7328	2.9269-4	4.0930-6

The leakage current in a bin is equal to the flux times the average cosine of the angle between the particle flight paths and the normal to the surface. If we assume the average angle is that for the line between the origin and the geometric center of the bin, we can estimate the leakage current for each bin. For bin (1,1) the cosine of the angle between the line from the source point to the center of the bin and the normal to the surface is approximately 0.9975. Thus the current and the flux in this bin for the present streaming calculation should be approximately equal. To be precise, within the limits of this approximation, the current through the bin should be 5.485×10^{-4}. The PFC calculation gives a result of $(5.504 \pm 0.048) \times 10^{-4}$. The ratio of the current to the flux calculated by PFC gives an estimate of the average cosine for particles scoring in the bin equal to 0.9967.

The cosine of the angle between the line from the source point to the center of bin (10,10) and the normal to the surface is 0.59708. Thus our analytic estimate for the current through this bin is 1.1763×10^{-4}. The PFC result for this current is $(1.170 \pm 0.022) \times 10^{-4}$. The agreement between these results is again well within statistics. The average cosine predicted by PFC for this bin is 0.5972. We conclude from this brief analysis of the results of the free-streaming calculation of flux and current in the Exercise 7.2 geometry that the geometry, source, and detectors are correctly placed and normalized.

Because there are 100 bins in which results are to be scored, the average number of source particles that contribute to the score per bin in a pure-scattering or particle-streaming problem is S/100, where S is the total number of start particles used in the calculation. Thus for $S = 10^6$ the average fractional standard deviation for the results obtained in each bin should be about 0.01. Because the number of scores will decrease for the cases with absorption, we conclude that when absorption is included in this calculation we will probably require more than 10^6 source particles to produce good-quality results in all detectors. Variance reduction might help in certain bins but, if we wish to score standard boundary-crossing detectors in all 100 bins simultaneously, variance reduction will not change the average number of scores per bin. As a result we use some brute force in these calculations and start 10^7 source particles per calculation. On this basis we are ready to complete the exercise.

Using the cross sections of Table 7.21, and 10^7 start particles per calculation, we obtain the flux results for the three parts of Exercise 7.2 shown in Figures 7.6, 7.7, and 7.8. In all cases the flux is a maximum at the center of the faces and a minimum at the corners. As with the free-streaming case, for the pure absorber we should be able to estimate the results at selected locations rather accurately. The magnitude of the flux on the surface in this case is given by

$$\phi(p) = \frac{Se^{-\left(1+\frac{1}{\mu}\right)\Sigma_t}}{4\pi t^2} \tag{7.4}$$

where S is the source strength, $t = t(p)$ is the distance the particle travels before escaping from the box at the point p, and $\mu = \mu(p)$ is the cosine of the angle between the particle trajectory and the normal to the surface. This equation assumes the two layers (sphere and box) are each one mfp thick. For the two points we examined in detail for the free-streaming case, bins (1,1) and (10,10), we have already determined t and μ for the mid-point of the bin. Thus from eqn 7.4, for the pure absorber case we estimate the flux in bin (1,1) to be 7.4197×10^{-5}. Multiplying by μ we get for the current 7.397×10^{-5}. Our PFC results for the flux and current in bin (1,1) are $(7.449 \pm 0.056) \times 10^{-5}$ and $(7.424 \pm 0.056) \times 10^{-5}$, respectively. These results correspond to an average cosine of 0.9967. For bin (10,10) we calculate a flux of 1.3578×10^{-6} and a current of 8.107×10^{-6}. The corresponding PFC results are $(1.342 \pm 0.031) \times 10^{-6}$ and $(8.021 \pm 0.183) \times 10^{-6}$, which give an average cosine of 0.5975. Obviously these results are all in agreement, within statistics, and help confirm that the calculation is correctly structured.

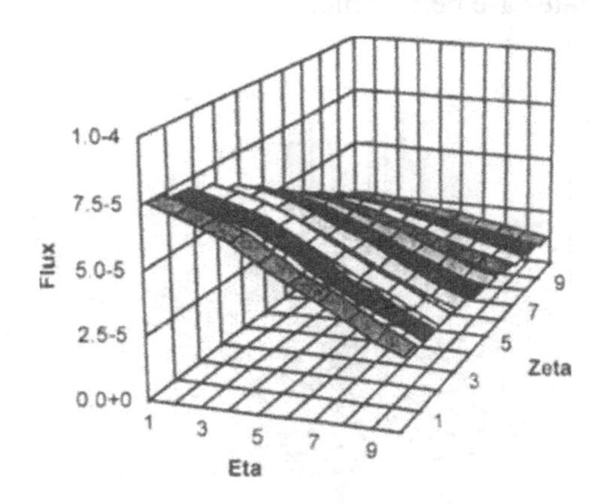

Figure 7.6. Surface Flux for Exercise 7.2a, Pure Absorbers

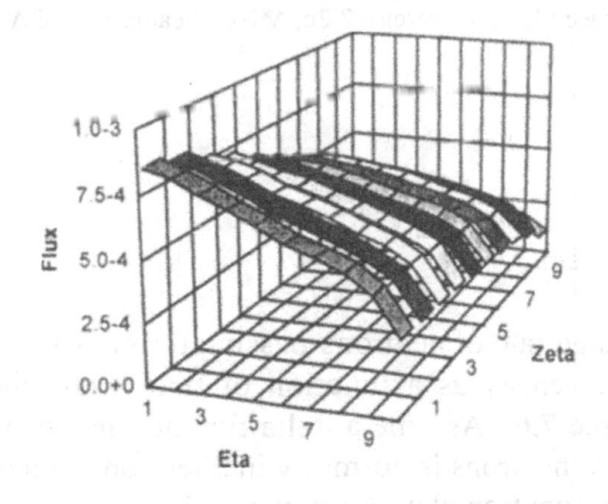

Figure 7.7. Surface Flux for Exercise 7.2b, Pure Scatterers

In each of Figures 7.6, 7.7, and 7.8 the front edge of the figure (for ζ cell 1) shows the surface flux falling steadily along the η axis from a maximum at the center of the face to a minimum at the edge. This fall should be matched by the fall along the η cell 1 results displayed against the left face of each figure. From the center of the edge, at (10,1), which is the closest point to the viewer in the figure, along the ζ axis direction to the right rear of the figures, the flux again decreases continuously. These results, which correspond to moving from the center of an edge to a corner, should be the same as the $\zeta = 10$ curve along the back face of the figures. The minimum value of the flux, which occurs at the point (10,10), corresponds to the cell at the corner of the box. Thus the general shape of the present results for the flux on the surface of the problem geometry is reasonable. Based on these considerations, and the comparisons for particle streaming and pure absorption, we believe we have obtained correct solutions to the problem and that the error estimates are believable.

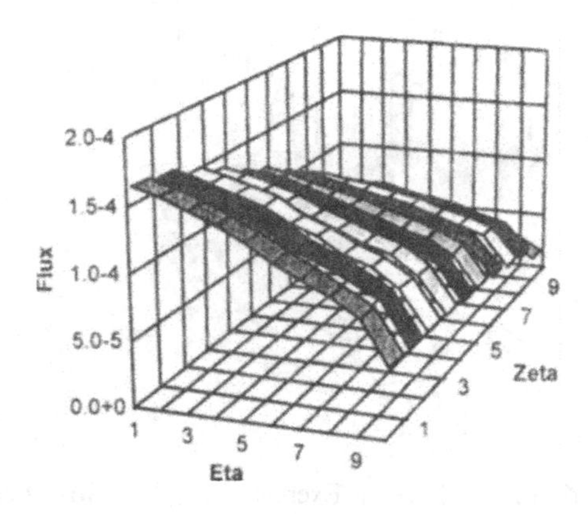

Figure 7.8. Surface Flux for Exercise 7.2c, Mixed Scatterers and Absorbers

Exercise 7.3

Statement of the problem

3. Calculate the leakage rate of neutrons passing through a 10-cm-thick slab of water of nominal density as a function of time. Use the water cross sections from Example 7.6. Assume a delta-function monoenergetic source at time zero of 5-MeV neutrons is normally incident on the upstream face of the slab. Account for neutron slowing down and absorption in the slab but assume all neutrons remain at thermal energy (0.025eV) when they reach this energy.

Solution

This problem is similar to Example 7.6 and many of the routines modified for solving that example can be used directly. One important change is the geometry. Because this exercise involves a parallel beam onto a slab we know that many of the source neutrons will be reflected from the slab. Therefore we expect the leakage through the slab to be significantly smaller than the leakage from the spherical geometry of Example 7.6 per source particle.

The desired leakage result does not depend on the lateral coordinates of either the incident or leaking neutrons and we can place our source anywhere on the upstream face of the slab. For example, we could use a pencil beam, several pencil beams, or a distributed parallel beam as our source. For convenience, we use a single pencil beam. Because the results do not depend on the lateral coordinates of the particles, we could eliminate tracking in the transverse directions as in Example 3.3. However, such elimination would require modifications to the three-dimensional tracking that we developed for PFC. Rather than do this we use a finite slab and assume the pencil beam is incident at the center of the upstream face.

By using finite lateral dimensions in our geometry we have the possibility of particles leaking through the sides of the slab. To verify that we do not lose an excessive number of neutrons by leakage through the lateral faces of the slab we tally such losses. The lateral dimensions we use should be sufficiently large that essentially all of the source neutrons that are not absorbed are either transmitted through or reflected from the slab. We use lateral X-Y dimensions of ±35 cm and assume the 10-cm thickness is in the Z-direction. The geometry used for this calculation is shown in Table 7.26.

Table 7.26. Geometry for Exercise 7.3

1	There is one RPP
-35. 35. -35. 35. 0. 10.	Coordinates of faces of the RPP
0	There are no SPHs
2	There are two zones
1	Zone one has one body
1	Body one defines zone 1
1	Zone two has one body
-2	Body two defines zone 2

Instead of a fission-spectrum source as in Example 7.6, this exercise calls for a monoenergetic source of neutrons. Therefore we expect a wave of uncollided neutrons to exist in the slab at early times, and for the downstream leakage to be zero until the wave reaches the downstream face of the slab. For the specified source energy of 5 MeV we know by eqn 7.44P that the initial neutron velocity is about 3.0925×10^9 cm/s. For the 10-cm-thick slab specified for this exercise, if we start source particles at time zero, the wave arrives at the downstream face at a time of about 3.234×10^{-9} sec. We set our time steps in such a manner as to measure the arrival of this wave of neutrons.

The subroutines that have been modified for use in this calculation are listed in Table 7.27. Most of these subroutines are identical to the corresponding subroutines used in Example 7.6. As in that example, the hydrogen and oxygen atom number densities are read by the version of subroutine 'Xsects' shown in Table 7.34P. These number densities are read from the file 'xsects.txt.' As in that example we use the nominal H and O number densities for water of 0.06692 and 0.03346 atoms/barn-cm. The modified subroutine 'Source' is shown in Table 7.28. This routine starts all source particles at the origin, traveling in the +Z direction, with an energy of 5 MeV. The age of the start neutrons is set to zero to correspond to a delta-function pulse in time at t = 0.

Table 7.27. Modified Subroutines for Exercise 7.3

Subroutine	Location
'Hydrogen'	Table 7.32P
'Oxygen'	Table 7.33P
'Xsects'	Table 7.34P
'Mxsec'	Table 7.35P
'Source'	Table 7.28
'Bdrx'	Table 7.29
'Stats'	Table 7.30
'Col'	Table 7.39P
'Isocol'	Table 7.40P

Table 7.28. Modified Subroutine 'Source' for Exercise 7.3

```
SUBROUTINE SOURCE                                                      !  1
REAL(8) x,y,z,u,v,w,xo,yo,zo,uo,vo,wo,wate,age,energ                  !  2
COMMON/PART/x,y,z,u,v,w,xo,yo,zo,uo,vo,wo,wate,age,energ,nzcur,newzn,ngroup !  3
pi=2.0d0*DACOS(0.0d0)          ! keep for more complicated source that needs pi   4
x=0.0d0;  y=0.0d0;  z=0.0d0 ! starts at origin                              5
nzcur=1                        ! assumes origin is in zone one                    6
u = 0.d0;  v = 0.d0;  w = 1.d0   ! Start particle in +Z direction
wate=1.0d0                     ! particle starts with a weight of one             8
age=0.0d0                      ! particle starts with an age of 0
energ=5.0d6                ! Start particle at 5 MeV
RETURN                                                                 !  9
END                                                                    ! 10
```

The modified subroutine 'Bdrx' is shown in Table 7.29. The variables 'refl' and 'xlat' are added to score the number of particles reflected from the slab and the number leaking through the lateral faces of the slab. Subroutine 'Bdrx' tests for reflection and lateral leakage and the appropriate variable is increased as required. Finally, if the particle leaks through the downstream face, the time interval during which the particle leaks is determined and the tally of particles for scoring the time-dependence of the leakage in that bin is incremented by the particle weight.

Table 7.29. Modified Subroutine 'Bdrx' for Exercise 7.3

```
SUBROUTINE BDRX                                                       !   1
REAL(8) delta,vel
REAL(8) sphcx(20),sphcy(20),sphcz(20),sphr(20),rpxmin(20),rpxmax(20),&  !   3
 rpymin(20),rpymax(20),rpzmin(20),rpzmax(20),bdin(40),bdout(40)       !   4
LOGICAL bdtest(40), bdhit(40)                                         !   5
COMMON/cg/sphcx,sphcy,sphcz,sphr,rpxmin,rpxmax,rpymin,rpymax,rpzmin,rpzmax,&  !  6
 bdin,bdout,nsph,nrpp,nbdy,nzones,nbz(40,41),nnext(40,40,40,2),bdtest,bdhit   !  7
REAL(8) x,y,z,u,v,w,xo,yo,zo,uo,vo,wo,wate,age,energ                 !   8
COMMON/PART/x,y,z,u,v,w,xo,yo,zo,uo,vo,wo,wate,age,energ,nzcur,newzn,ngroup  !  9
REAL(8) dmfp,dtr,xsec,dcur                                            !  10
COMMON/TRACK/dmfp,dtr,xsec,dcur                                       !  11
REAL(8) bscore(22),bsumsq(22),btime(22),dtime(22),bpart(22)
COMMON/STAT/bscore,bsumsq,btime,dtime,bpart
COMMON refl, xlat,xleak       ! add vbls for reflection, lateral leakage, and leaks
vel=1.38318d6*DSQRT(energ)    ! velocity in cm per sec for energy in eV
delta=dtr-dcur                      ! delta--distance traveled to reach boundary  12
dcur=dtr                            ! update current distance traveled            13
dmfp=dmfp-delta*xsec                ! subtract current distance in mfp from dmfp  14
x=x+u*delta;  y=y+v*delta;  z=z+w*delta     ! update position                     15
age=age+delta/vel   ! update age of particle
IF(age.GE.1.0d-2)THEN  ! check to see if particle has exceeded age limit
  nzcur=-1             ! exceeded age, set flag to kill
  RETURN
ENDIF
nzcur=newzn                         ! change identifier of current zone           16
if(nzcur.LT.nzones)RETURN           ! if not at outer zone then return            17
nzcur=-1                            ! set nzcur=-1 if in outer zone (for escape)  18
IF(z.GT.9.9999) THEN   ! if leaking particle has escaped thru downstream face
  xleak=xleak+wate
  DO i=1,22
    IF(age.LE.btime(i))THEN       !  if leakage occurs in this time interval
      bpart(i)=bpart(i)+wate      !  then score and return
      RETURN
    END IF
  END DO
  RETURN
ENDIF
IF(z.LT.0.0001) THEN   !  if reflection
  refl = refl + wate
  RETURN
ENDIF
xlat = xlat + wate     ! must be lateral leakage
RETURN
END                                                                   !  20
```

The modified version of subroutine 'Stats' is shown in Table 7.30. The boundaries for the time bins used in this calculation are defined at entry point 'StatOne.' In Example 7.6 we use 19 time bins and the earliest time bin boundary is set to 10^{-8} sec following the source pulse. In this problem, to detect the source neutron wave arrival time we add time boundaries of 2, 3, and 4 x 10^{-9} sec while keeping the 19 bins used in Example 7.6. Thus the number of time steps is increased from 20 to 22, and the tally variables are dimensioned accordingly.

Table 7.30. Modified Subroutine 'Stats' for Exercise 7.3

```
 SUBROUTINE Stats                                                           !     1
      REAL(8) bscore(22),bsumsq(22),btime(22),dtime(22),bpart(22)
      COMMON/STAT/bscore,bsumsq,btime,dtime,bpart
      COMMON/IN/npart,nbatch                                                !     4
      REAL(8) tmp,tmp1,tmp2,var,stdev                                       !     7
      COMMON refl, xlat, xleak          !  vbls for refl and lateral leakage
 ENTRY StatOne          !  entry point to initialize arrays for complete problem  8
      refl = 0. ; xlat = 0.; xleak=0.
      bscore=0.0d0;  bsumsq=0.0d0
      btime(1)=3.0d-9;  dtime(1) = btime(1) ! changed time steps, not from example 7.6
      btime(2)=4.0d-9;  dtime(2) = 1.0d-9
      btime(3)=5.0d-9;  dtime(3) = dtime(2)
      btime(4)=1.0e-8;  dtime(4)=btime(4)-btime(3)
      tmp=DEXP(DLOG(1.d6)/18.0d0)       !  to create 18 more log spaced time banks
      DO i=5,22
        btime(i)=btime(i-1)*tmp         !  max age for particles in this bank
         dtime(i)=btime(i)-btime(i-1)   !  delta time for particles in this bank
      END DO
 RETURN                                                                     !    10
 ENTRY StatLp           !  entry point to initialize arrays for a particle        11
      bpart=0.0d0
 RETURN                                                                     !    13
 ENTRY StatELp          !  entry point to store scores for a particle             14
     DO 70 i=1,22
       bscore(i)=bscore(i)+bpart(i)     !  add particle score to total
        bsumsq(i)=bsumsq(i)+bpart(i)**2 !  sum for variance calculation
 70  CONTINUE
 RETURN                                                                     !    19
 ENTRY StatEnd          !  entry point to calculate and print results             20
      OPEN(UNIT=18,FILE='output2.txt')
      tmp=DFLOAT(npart)                                                     !    21
      DO 90 i=1,22
        var=bsumsq(i)/tmp-(bscore(i)/tmp)**2       !  variance
        stdev=DSQRT(var)                           !  standard deviation
        WRITE(18,10)i,btime(i),bscore(i)/tmp/dtime(i),stdev/DSQRT(tmp)/dtime(i)
        WRITE(16,10)i,btime(i),bscore(i)/tmp/dtime(i),stdev/DSQRT(tmp)/dtime(i)
 90   CONTINUE
 10   FORMAT(i3,1p3e12.3)
      WRITE(16,1) refl, xlat, xleak !  write number of reflected, lateral leakage and
end leaks
  1   FORMAT(/,' Weight of reflected particles = ',1pe12.3,/&
             ' Weight of lateral leak parts   = ',1pe12.3,/&
             ' Weight of transmitted parts    = ',1pe12.3)
 RETURN                                                                     !    28
      END                                                                   !    29
```

The results obtained from running 10^6 start particles in this program are shown in Table 7.31. We see that, as expected, no neutrons escape through the downstream face of the slab during time step one ($0 - 3 \times 10^{-9}$ sec). The pulse of uncollided source neutrons arrives between $3 - 4 \times 10^{-9}$ sec, after which the leakage rate falls rapidly. Finally, by the end of time step 20, at 2.154×10^{-3} sec, all neutrons have escaped from the system and the leakage rate falls to zero. The downstream leakage rate is plotted as a function of time in Figure 7.9. Except for the arrival pulse and the time of die away, the time-dependence of the leakage is similar to that of Example 7.6, shown in Figure 7.8P. Because the present system is a slab, with leakage through both faces, and because the slab is much thinner than the system in Example

7.6 (a 10-cm-thick slab vs. a 30-cm-diameter sphere), the neutrons die away sooner in this calculation than in the example calculation.

Table 7.31. Results of Exercise 7.3 with 10^6 Source Particles

Time step	Time (s)	Leakage (n/s)	Sigma
1	3.000E-09	0.000E+00	0.000E+00
2	4.000E-09	3.934E+08	4.885E+05
3	5.000E-09	5.739E+07	2.326E+05
4	1.000E-08	1.698E+07	5.575E+04
5	2.154E-08	3.169E+06	1.626E+04
6	4.642E-08	5.864E+05	4.820E+03
7	1.000E-07	1.500E+05	1.666E+03
8	2.154E-07	5.658E+04	6.976E+02
9	4.642E-07	2.334E+04	3.053E+02
10	1.000E-06	1.024E+04	1.377E+02
11	2.154E-06	4.341E+03	6.100E+01
12	4.642E-06	1.806E+03	2.673E+01
13	1.000E-05	7.967E+02	1.199E+01
14	2.154E-05	5.206E+02	6.459E+00
15	4.642E-05	3.965E+02	3.669E+00
16	1.000E-04	2.615E+02	1.848E+00
17	2.154E-04	1.178E+02	7.024E-01
18	4.642E-04	2.465E+01	1.557E-01
19	1.000E-03	1.154E+00	1.223E-02
20	2.154E-03	3.266E-03	1.243E-04
21	4.642E-03	0.000E+00	0.000E+00
22	1.000E-02	0.000E+00	0.000E+00

From subroutines 'Hydrogen' and 'Oxygen' we can obtain the microscopic cross sections used in this calculation for H and O at 5 MeV; viz., 1.5 and 0.9 barns, respectively. Thus, using the number densities given above, the total macroscopic cross section for a 5-MeV neutron in our water slab is about $\Sigma_t = 0.1305\ \text{cm}^{-1}$. This corresponds to a mean free path of 7.66 cm; i.e., our slab is about 1.3 mfp thick. Because the source is normally incident on the slab, 27.1% of the incident particles should penetrate the slab without suffering a collision, and the leakage rate in our second time step should be not less than this number divided by the width of the time step. The latter is 10^{-9} sec and thus we estimate the minimum leakage rate in the second time step to be 2.71×10^8 n/s. The calculated leakage rate for this time step, from Table 7.26, is 3.93×10^8 n/s. We conclude that, in addition to the uncollided source neutrons, a number of source neutrons are scattered through a small angle, as measured in the laboratory coordinates, and also leak during this second time step.

The weight of the reflected particles is calculated to be 2.119×10^5 out of a total source weight of 10^6. Thus the probability of reflection is about 21%. The weight of particles escaping through the lateral faces of the slab is found to be 517.6. The probability that a source neutron will escape through the side of the slab is about 5×10^{-4}. This is sufficiently small that we do not need to correct for systematic bias in our results, and we conclude that our lateral slab dimensions of ±35 cm are adequate for the present calculation.

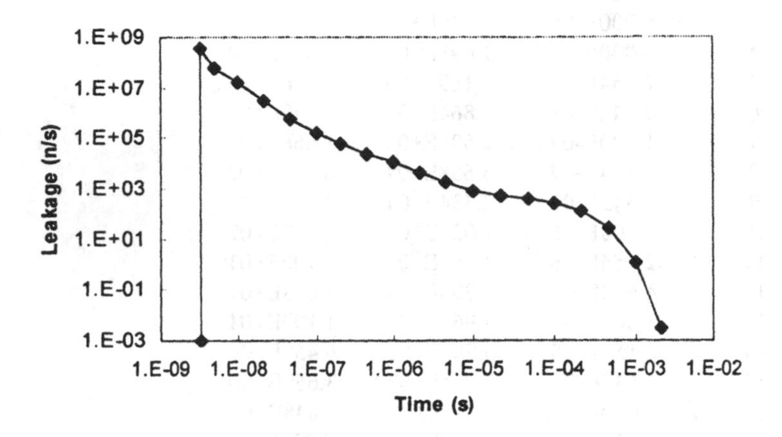

Figure 7.9. Time-Dependent Leakage for Exercise 7.3

Chapter 8

Nuclear Criticality Calculations with Monte Carlo

Exercise 8.1

Statement of the problem

1. Find thc value of c for which two spheres of homogeneous fissile material, each 5 mean free paths in diameter and separated by one-fifth their diameter, would be exactly critical. Assume the neutrons are monoenergetic and scattering is isotropic in the laboratory system. The spheres are surrounded by void.

Solution – Generation Method

This problem is similar to that of Examples 8.1 and 8.2 and we use those solutions as a starting point for solving the current exercise. Because the geometry is defined in terms of the mfp of the particles, we are free to choose any size we wish for the spheres provided we adjust the total cross section to meet the problem specification. The geometry we choose to use for this calculation is shown in Table 8.1. We define two spheres, of radius 2.5 cm each, the centers of which are located on the X axis at x = ± 3 cm. Thus at their point of closest approach the spheres are separated by 1 cm. To make the 5-cm diameter equal to five mfp we set the total cross section in the spheres to one. The region between the two spheres is a void and thus neutrons can pass freely from one sphere to the other. The cross sections used in this calculation are defined as in Example 8.1 except here we have two zones of fissile material. Both zones are assumed to contain the same material so the value of c, the non-absorption probability, must be the same in both spheres.

Table 8.1. Geometry for Exercise 8.1

0	No RPPs
3	Three SPHs
-3. 0. 0. 2.5	Center and radius of SPH 1
3. 0. 0. 2.5	Center and radius of SPH 2
0. 0. 0. 25.	Center and radius of SPH 3
4	Number of zones
1	No. bodies in zone 1
2	Definition of zone 1
1	No. bodies in zone 2
1	Definition of zone 2
3	No. bodies in zone 3
3 -1 -2	Definition of zone 3
1	No. bodies in zone 4
-3	Definition of zone 4

Let us begin by using the generation and total subsequent population methods to find the critical value of c. The subroutines modified to solve this exercise are listed in Table 8.2. All but one of the subroutines are identical to those used in Example 8.1. The one modified specifically for this exercise, subroutine 'Source,' is shown in Table 8.3. The modified subroutine 'Source' allows three different first-generation source distributions for this split-core calculation: a point source at the center of the x > 0 sphere, two equally probable point sources located at the centers of both spheres, and a source distributed uniformly over both spheres. The updates for the two source distributions that are not being used in a calculation must be commented out. In addition, entry point 'Sourc3' is modified to define the correct zone for a source particle. From Table 8.1 we see the interior of the x > 0 sphere is defined as zone 1 and the interior of the x < 0 sphere is defined as zone 2. Therefore the modified 'Sourc3' tests the x value of each start particle and assigns the correct zone.

Table 8.2. Modified Subroutines for Exercise 8.1 Generation Calculation

Subroutine	Location
'Walk'	Table 8.3P
'Source'	Table 8.3
'Col'	Table 8.5P
'Stats'	Table 8.6P
'Input'	Table 8.7P

We found in Example 8.1 that the normal-mode flux is achieved in the geometry of that problem after about 10 generations. This result is used in setting the variable 'ntransient' in subroutine 'Walk.' We have no reason to believe that the equilibrium-mode flux will be achieved in this split-core problem in the same number of generations. Therefore we use the three

Table 8.3. Modified Subroutine 'Source' for Exercise 8.1 Generation Calculation

```
      SUBROUTINE SOURCE                                                !    1
      REAL(8) FLTRN,r1,r2,tmp,radius,costh,sinth,phi,pi
      COMMON/IN/npart,nbatch
      REAL(8) x,y,z,u,v,w,xo,yo,zo,uo,vo,wo,wate,age,energ             !    2
      COMMON/PART/x,y,z,u,v,w,xo,yo,zo,uo,vo,wo,wate,age,energ,nzcur,newzn,ngroup  !  3
      REAL(8) sphcx(20),sphcy(20),sphcz(20),sphr(20),rpxmin(20),rpxmax(20),&
       rpymin(20),rpymax(20),rpzmin(20),rpzmax(20),bdin(40),bdout(40)
      LOGICAL bdtest(40), bdhit(40)
      COMMON/cg/sphcx,sphcy,sphcz,sphr,rpxmin,rpxmax,rpymin,rpymax,rpzmin,rpzmax,&
       bdin,bdout,nsph,nrpp,nbdy,nzones,nbz(40,41),nnext(40,40,40,2),bdtest,bdhit
      REAL(8) xbank(20000,2),ybank(20000,2),zbank(20000,2),wbank(20000)
      COMMON/BANK/xbank,ybank,zbank,wbank,nsave
      pi=2.0d0*DACOS(0.0d0)        ! keep for more complicated source that needs pi   4
      DO 100 i=1,npart
!  point source at center of sphere 2, x > 0
!       xbank(i,2)=3.0d0;  ybank(i,2)=0.0d0;  zbank(i,2)=0.0d0
!  point source at center of each sphere, equal prob -
!  retain previous updates and add:
!        IF(fltrn().LT.0.5) xbank(i,2) = -3.d0
!  uniform distribution over both spheres
        radius = 2.5d0*fltrn()**(1.0d0/3.0d0)
        costh = -1.d0+2.d0*fltrn();  sinth = DSQRT(1.-costh**2);  phi = 2.d0*pi*fltrn()
        zbank(i,2) = radius*costh;  xbank(i,2) = 3.d0+radius*sinth*DCOS(phi)
        ybank(i,2) = radius*sinth*DSIN(phi)
        IF(fltrn().LT.0.5) xbank(i,2) = -xbank(i,2)
        wbank(i)=1.0d0
  100 CONTINUE
      nsave=npart
      RETURN
      ENTRY SOURC2
      initial=npart;  npart=0
      DO 200 i=1,nsave
  150 CONTINUE
      IF(wbank(i).GE.1.0d0)THEN
        npart=npart+1
        xbank(npart,1)=xbank(i,2); ybank(npart,1)=ybank(i,2); zbank(npart,1)=zbank(i,2)
        wbank(i)=wbank(i)-1.0d0;  GOTO 150
      ELSE
        IF(wbank(i).GE.FLTRN())THEN
          npart=npart+1
          xbank(npart,1)=xbank(i,2);ybank(npart,1)=ybank(i,2);zbank(npart,1)=zbank(i,2)
        ENDIF
      ENDIF
  200 CONTINUE
      IF(npart.GT.initial)npart=initial       !  keeps particle numbers from increasing
      nsave=0
      IF (npart.LE.0) THEN
        write(*,*)' The number of particles decreased to zero.'
        STOP
      ENDIF
      IF(npart.LT.initial)THEN                !  keeps particle numbers from decreasing
        ntemp=initial-npart
        DO i=1,ntemp
          npart=npart+1
          xbank(npart,1)=xbank(i,1);ybank(npart,1)=ybank(i,1);zbank(npart,1)=zbank(i,1)
        ENDDO
      ENDIF
      RETURN
      ENTRY SOURC3(n)
      x=xbank(n,1);  y=ybank(n,1);  z=zbank(n,1);  nzcur = 1;  wate=1.0d0
      IF(x.lt.0.) nzcur = 2  !  two source zones, zone 1 for x > 0, zone 2 for x < 0
      CALL ISOOUT                    !  direction chosen isotropically                 7
      RETURN;  END
```

starting source distributions defined above and examine the batch k values to determine the number of generations we should calculate before beginning to make estimates of k. The 'write' statement that is commented out under entry point 'StatEBatch' of subroutine 'Stats' must be activated and 'ntransient' set to zero in 'Walk' to record these batch k values.

Running the modified PFC with c = 1.17 and 10^4 start particles per generation for 20 generations, and using each of the three source distributions given above in turn, gives the results shown in Figure 8.1. The single- and double-point-source distributions, which give almost identical results, overestimate k initially because they place source neutrons at locations of high importance. The third distribution, a uniform source over both spheres, underestimates k initially because neutrons are started with equal probability in locations of low importance and in locations of high importance. However, all three source distributions converge to the normal mode flux within ten generations. Therefore we retain the setting of 'ntransient' equal to ten in subsequent calculations for this exercise.

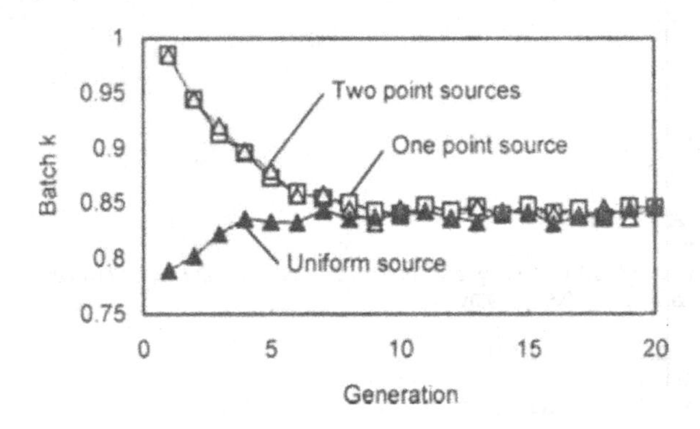

Figure 8.1. Values of k for Early Generations

From Figure 8.1 we see that c = 1.17 produces a system that is subcritical. Therefore we calculate k for values of c between 1.2 and 1.35. The results of such calculations, using the uniform distribution for our initial source, 10^4 source particles per generation, 'ntransient' = 10, and 50 generations following 'ntransient' initial generations, are shown in Table 8.4. These data are plotted in Figure 8.2. Both the generation and total-subsequent-population estimates of k are shown. As before the two results for k are identical but the standard deviations are different. In Example 8.1 it is shown that the estimate of the standard deviation based on the total-subsequent-population result is more accurate than that based on the generation method. Using linear regression on the results of Table 8.4 we find

$$k \approx 0.78911c - 4.468 \times 10^{-3}$$

Thus we estimate that the system is critical for $c \approx 1.2729$. A set of long computer runs gives a result of $c \approx 1.27435$.

Table 8.4. Calculated Generation and Total Subsequent Population k, Exercise 8.1

c	gen k	sigma	tot sub pop k	sigma
1.2	0.94259	6.40E-04	0.94259	6.97E-04
1.25	0.981505	7.50E-04	0.981505	7.26E-04
1.3	1.021826	7.79E-04	1.021826	7.54E-04
1.35	1.060668	7.77E-04	1.060668	7.83E-04

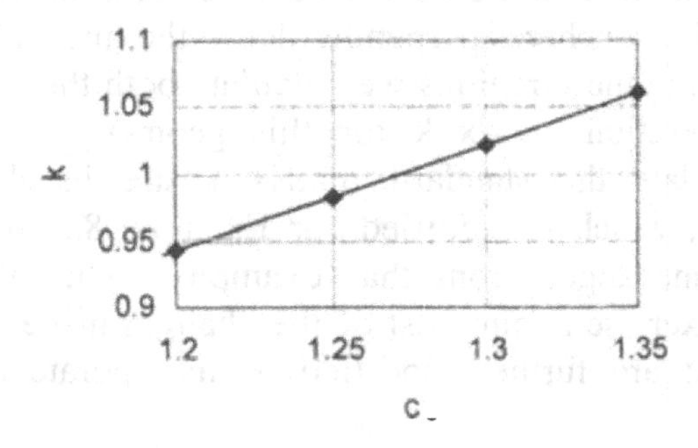

Figure 8.2. k as a function of c, Exercise 8.1

Solution – Standard Matrix Method

The split-core system of Exercise 8.1 is seemingly designed for using matrix k to estimate the neutron multiplication of the system. However, the standard matrix method requires the user to define a source distribution in each of the matrix regions. In this problem we do not have spherical symmetry as we do in Example 8.2. In that example we use concentric spherical shells as our matrix regions and assume a uniform source distribution in each region. By making the spherical shells thin we can be certain that this assumption is reasonable. Our matrix results are in good agreement with the generation results, indicating that the uniform-source assumption for the matrix calculation is acceptable for that calculation.

In the present problem the equilibrium flux again varies with radius in each sphere, but the portions of the spheres nearest the origin are more important, and have a higher equilibrium flux, than the portions farthest from the origin. That is, in this problem the flux varies with position along the line through the centers of the two spheres as well as with the distance from this line. If we divide both spheres into regions that are symmetric across the Y-Z plane (i.e., the plane perpendicular to the line between centers), we know that the resulting fission matrix will be symmetric. However, this does not help to define the approximate initial flux distribution because no matter how thin we make such regions we will never be able to approximate the equilibrium flux by a uniform distribution within each region. The flux will always vary with distance from the axis, a dependence that a uniform source distribution neglects. We therefore expect that using a uniform source distribution in these regions will produce incorrect results from a standard matrix calculation. On the other hand, the combined generation and matrix method, which allows an equilibrium flux distribution to be established in the regions before making the matrix estimate of k, should work well for these proposed regions.

To verify this contention we define regions for a matrix k calculation using slices through the spheres perpendicular to the line through the centers of the spheres. Using these regions we calculate both the standard matrix k and the multi-generation matrix k for this geometry. The subroutines modified to calculate the standard matrix k are listed in Table 8.5. Subroutine 'Walk', which is modified for Example 8.2 to include a loop over regions, is unchanged from that example. The other subroutines modified for this exercise retain most of the changes made for the Example 8.2 calculation but are further modified to incorporate the new region definitions.

Table 8.5. Subroutines Modified for Basic Matrix k Calculation, Exercise 8.1

Subroutine	Location
'Walk'	Table 8.14P
'Input'	Table 8.6
'Source'	Table 8.7
'Col'	Table 8.8
'Stats'	Table 8.9
'keff'	Table 8.19P

Subroutine 'Input' is shown in Table 8.6. This version of the routine is similar to that of Table 8.15P except that here, by symmetry, we would like the number of regions in each sphere to be the same. Therefore we modify subroutine 'Input' to include a user-specified number of regions in each sphere, and then double this input value to obtain the total number of regions. This doubling guarantees an even number of regions in the calculation. We retain a maximum of 20 regions, thus allowing up to ten regions in each sphere. The regions in each sphere are selected symmetrically about the center of the sphere. This is shown in Figure 8.3.

Table 8.6. Modified Subroutine 'Input' for Exercise 8.1 Matrix k

```
      SUBROUTINE INPUT                                        !    1
      COMMON/IN/npart,nregions,iter
      WRITE(*,10)
   10 FORMAT(' Enter the number of regions in each sph. (max of 10) ')
      READ(*,*)nregions
      IF(nregions.LE.0.OR.nregions.GT.10)nregions=10
      WRITE(16,*)'Number of regions/sphere = ',nregions
      nregions = 2*nregions  !  double regions for two spheres
      WRITE(*,11)
   11 FORMAT(' Enter the number of matrix iterations. ')
      READ(*,*)iter
      WRITE(16,*)' Number of matrix iterations = ',iter
      WRITE(*,12)                                             !    3
   12 FORMAT(' Enter the number of particles/region.')
      READ(*,*)npart                                          !    5
      WRITE(16,*)'Number of particles/region = ',npart
      WRITE(*,13)                                             !    7
   13 FORMAT(' Enter the random number seed, an integer. ')   !    8
      READ(*,*)iseed                                          !    9
      IF(iseed.LT.1)iseed=1                                   !   10
      IF(iseed.GE.2147483647)iseed=2147483646                 !   11
      WRITE(16,*)' Starting random number seed is ',iseed     !   12
      CALL rndin(iseed)                                       !   13
      RETURN                                                  !   15
      END                                                     !   16
```

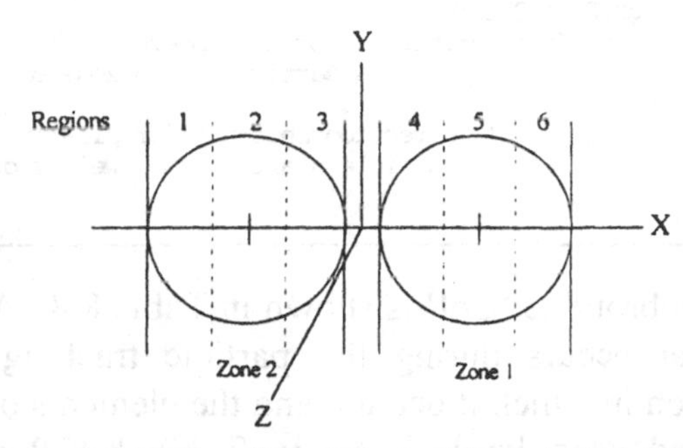

Figure 8.3. Region Definition for Matrix k Calculation

Subroutine 'Source,' shown in Table 8.7, is modified to produce a uniform source distribution over each of the matrix regions. This is done by finding the maximum radius of each region in the direction perpendicular to the line between the centers of the spheres, and using the rejection technique to obtain source points within the region. The geometry used for the matrix calculation is the same as that used in the generation calculation; i.e., the centers of the spheres are located at x = ± 3 cm and the spheres have radii of 2.5 cm. These dimensions are coded into the routine. The regions are numbered as indicated in Figure 8.3, starting with the left boundary of region 1, at x = -5.5, and cycling through the regions with increasing x values. The widths of the regions are set by the number of regions used in the calculation, as coded in subroutine 'Stats.'

Table 8.7. Modified Subroutine 'Source'

```
      SUBROUTINE SOURCE                                                  !  1
      REAL(8) x,y,z,u,v,w,xo,yo,zo,uo,vo,wo,wate,age,energ               !  2
      COMMON/PART/x,y,z,u,v,w,xo,yo,zo,uo,vo,wo,wate,age,energ,nzcur,newzn,ngroup  !  3
      REAL(8) aij(20,20),aijsq(20,20)
      COMMON/STAT/aij,aijsq
      COMMON/EXAMPLE1/nreg ; COMMON/exercise1/deltax,nregx
      REAL(8) radius,fltrn
      COMMON/IN/npart,nregions,iter
!  source for region nreg - start with x < 0
!  hard coded for 5 cm dia spheres centered at x = + or - 3.
!  define x bounds for region nreg - assumes sphere at origin
      IF(nreg.LE.nregx)THEN
        xmin = FLOAT(nreg-1)*deltax-2.5
        xshift = -3. !   start with sphere 1 = x<0, then shift to 2 = x>0
      ELSE
        xmin = FLOAT(nreg-nregx-1)*deltax-2.5
        xshift = 3.
      ENDIF
      xmax = xmin+deltax
      amin = ABS(xmin) ! coding picks a point in nreg assuming sphere center at origin
      IF(ABS(xmax).lt.amin)amin = xmax
      t = SQRT(2.5**2-amin**2)
      IF(xmin-0.00001.lt.0..and.xmax+0.00001.gt.0.)t = 2.5
 2    x = xmin + fltrn()*deltax    !  find source point using rejection
      y = t*fltrn()
      z = t*fltrn()
      IF(x**2+y**2+z**2.gt.2.5**2)GOTO 2
      x = x+xshift        ! shift to either x<0 or x>0 sphere
      nzcur=1                      !  assumes particle is in zone one
      IF(x.lt.0.)nzcur = 2
      CALL ISOOUT                  !  direction chosen isotropically        7
      wate=1.0d0                   !  particle starts with a weight of one  8
      RETURN                                                             !  9
      END                                                                !  10
```

The modified subroutine 'Col' is shown in Table 8.8. As in Example 8.2 each collision that occurs during the particle tracking is examined to determine the region in which it occurs, and the elements of the matrix array 'aij' are assembled accordingly. As in Table 8.17P all collisions are assumed to result in the absorption of the incoming particle.

The modified subroutine 'Stats' is shown in Table 8.9. At entry 'Statone' the widths of the regions, 'deltax,' and the number of regions in each sphere, 'nregx,' are calculated and the tally arrays are set to zero. At the final entry point, 'StatEnd,' the standard deviations for the matrix elements are calculated and subroutine 'keff' is called. This subroutine, which is identical to that used in Example 8.2, performs the iteration of eqns 8.10P and 8.11P to obtain the multiplication factor for the system.

Running PFC as modified for this matrix k calculation, using c = 1.27435, three regions per sphere, 25 matrix iterations, and 10^4 source particles per region, we obtain the 6 × 6 fission matrix shown in Table 8.10. The elements of this matrix should be symmetric about the center point; i.e., we should have $a_{ij} = a_{7-j,7-i}$. From the table we see that this is indeed the case, within statistics (not shown). Thus the first and last elements of the matrix are approximately equal, $a_{1,1} = 0.702 \approx a_{6,6} = 0.699$. Likewise $a_{1,2} \approx$

$a_{6,5}$, etc. This calculation produced no fission events in elements $a_{1,5}$ and $a_{1,6}$, or in $a_{6,1}$ and $a_{6,2}$; i.e., the contribution to fissions at the outer edge of one sphere from fission neutrons born at the outer edge of the other sphere is small in this unreflected system. We conclude that the results of our simple matrix scheme are internally consistent, and if our assumption of a uniform source distribution in each matrix region is reasonable we should expect to obtain an accurate estimate of k.

Table 8.8. Modified Subroutine 'Col', Exercise 8.1 Matrix k

```
      SUBROUTINE COL                                                    !   1
      REAL(8) FLTRN,delta                                               !   2
      REAL(8) x,y,z,u,v,w,xo,yo,zo,uo,vo,wo,wate,age,energ              !   3
      COMMON/PART/x,y,z,u,v,w,xo,yo,zo,uo,vo,wo,wate,age,energ,nzcur,newzn,ngroup  ! 4
      REAL(8) dmfp,dtr,xsec,dcur                                        !   5
      COMMON/TRACK/dmfp,dtr,xsec,dcur                                   !   6
      REAL(8) sigt(20),c(20);  ! dimensions allow up to 20 different media   7
      COMMON/GEOM/sigt,c; !sigt is total cross section, c is non-absorption prob  8
      REAL(8) aij(20,20),aijsq(20,20)
      COMMON/STAT/aij,aijsq
      COMMON/IN/npart,nregions,iter
      COMMON/EXAMPLE1/nreg ; COMMON/exercise1/deltax,nregx
      delta=dmfp/xsec                 ! distance traveled to collision     9
      dtr=dcur+delta                  ! update total distance traveled    10
      x=xo+u*dtr;  y=yo+v*dtr;  z=zo+w*dtr      ! update position         11
!   find index
      IF(x.LT.0.)THEN           ! sph 1, x<0
        index = 0
        DO 2 i = 1,nregx
          index = index+1
          IF(x.lt.FLOAT(i)*deltax-5.5) GOTO 3
 2      CONTINUE
        WRITE(*,*)i,nregx,index,deltax,x
        STOP 2
      ELSE                      ! sphere 2, x>0
        index = nregx
        DO 4 i = 1, nregx
          index = index+1
          IF(x.lt.FLOAT(i)*deltax+0.5) GOTO 3
 4      CONTINUE
        WRITE(*,*)i,nregx,index,deltax,x
        STOP 2
      ENDIF
 3    CONTINUE
      aij(nreg,index)=aij(nreg,index)+c(nzcur)    ! scores all outgoing neutrons
      aijsq(nreg,index)=aijsq(nreg,index)+c(nzcur)**2
      nzcur=-1                        ! all incoming neutrons are absorbed
      RETURN                                                            !  21
      END                                                               !  22
```

Using c = 1.27435, 10^5 source particles per region, ten matrix regions per sphere, and 25 iterations of eqns 8.10P and 8.11P to estimate k, for the standard matrix calculation we obtain a result of k = 0.9465 ± 0.0007. From the generation method results given above we know the system should be approximately critical for this value of c. Additional calculations can quickly confirm that this region model with a uniform source distribution does not converge to the generation results regardless of the number of regions used. This is because the normal mode flux distribution is never uniform across the regions selected, regardless of how thin such regions are chosen to be.

Table 8.9. Modified Subroutine 'Stats', Exercise 8.1 Matrix k

```
 SUBROUTINE Stats                                                          !    1
     REAL(8) aij(20,20),aijsq(20,20)
     COMMON/STAT/aij,aijsq
     COMMON/IN/npart,nregions,iter
     REAL(8) x,y,z,u,v,w,xo,yo,zo,uo,vo,wo,wate,age,energ                  !    5
     COMMON/PART/x,y,z,u,v,w,xo,yo,zo,uo,vo,wo,wate,age,energ,nzcur,newzn,ngroup  !  6
     REAL(8) tmp,tmp1,tmp2,var,stdev,xkeff
     REAL(8) sphcx(20),sphcy(20),sphcz(20),sphr(20),rpxmin(20),rpxmax(20),&
      rpymin(20),rpymax(20),rpzmin(20),rpzmax(20),bdin(40),bdout(40)
     LOGICAL bdtest(40), bdhit(40)
     COMMON/cg/sphcx,sphcy,sphcz,sphr,rpxmin,rpxmax,rpymin,rpymax,rpzmin,rpzmax,&
      bdin,bdout,nsph,nrpp,nbdy,nzones,nbz(40,41),nnext(40,40,40,2),bdtest,bdhit
     COMMON/EXAMPLE1/nreg ; COMMON/exercise1/deltax,nregx
 ENTRY StatOne          ! entry point to initialize arrays for complete problem  8
!  define nregion matrix regions as slices through the spheres
!  in Y-X plane but assume sphere centered at origin
     nregx = nregions/2
     deltax = 5./FLOAT(nregx)
!  write(*,*)nregx, nregions, deltax
     aij=0.0d0;  aijsq=0.0d0
 RETURN                                                                    !   10
 ENTRY StatLp           ! entry point to initialize arrays for a particle  11
 RETURN                                                                    !   13
 ENTRY StatELp          ! entry point to store scores for a particle       14
 RETURN                                                                    !   19
 ENTRY StatEnd          ! entry point to calculate and print results       20
     tmp=DFLOAT(npart)                                                     !   21
     DO i=1,nregions
       DO j=1,nregions
         var=aijsq(i,j)/tmp-(aij(i,j)/tmp)**2
         aij(i,j)=aij(i,j)/tmp
         aijsq(i,j)=var/tmp
       END DO
     END DO
!  write(16,1)((aij(i,j),i=1,nregions),j=1,nregions)
! 1format(/,' Fission matrix',/,(1p6e11.2))
!  write(16,2)((aijsq(i,j),i=1,nregions),j=1,nregions)
! 2format(/,' Standard deviations',/,(1p6e11.2))
     CALL keff(xkeff,var)
     WRITE(16,100)xkeff,DSQRT(var)
  100 FORMAT('  k-effective = ',1pe14.5,' +- ',1pe14.5)
 RETURN                                                                    !   28
     END                                                                   !   29
```

Table 8.10. Example Fission Matrix for Exercise 8.1, 'nregions' = 3

7.02E-01	8.54E-02	5.99E-03	5.10E-04	0.00E+00	0.00E+00
1.65E-01	7.85E-01	1.64E-01	8.67E-03	1.53E-03	3.82E-04
8.03E-03	8.81E-02	7.14E-01	3.38E-02	6.37E-03	7.65E-04
5.10E-04	3.70E-03	3.21E-02	6.99E-01	9.39E-02	6.88E-03
2.55E-04	1.40E-03	9.81E-03	1.67E-01	7.97E-01	1.68E-01
0.00E+00	0.00E+00	5.10E-04	8.28E-03	8.53E-02	6.99E-01

Solution – Matrix by Generation

We postulate that using several generations to obtain a source distribution that approximates that of the equilibrium flux in the system will allow us to obtain the correct system multiplication using a matrix calculation, even if the flux in a matrix region varies significantly with distance from the line of symmetry. To verify this we use the region definitions of Figure 8.3 in a matrix-by-generation calculation. The subroutines modified to calculate the system multiplication using the matrix-by-generation method are shown in Table 8.11. The major differences between these routines and those modified for Example 8.3 are related to the definition of matrix regions used in the present problem.

Table 8.11. Modified Subroutines for Exercise 8.1, Matrix by Generations

Subroutine	Location
'Walk'	Table 8.23P
'Input'	Table 8.12
'Source'	Table 8.13
'Col'	Table 8.14
'Stats'	Table 8.15
'keff'	Table 8.28P

The modified subroutine 'Input' is shown in Table 8.12. As in Table 8.24P, this version of subroutine 'Input' allows the user to specify both batch and matrix data. This routine is similar to that shown in Table 8.6 and again the number of matrix regions is defined such that there is an equal number in each of the two spheres, with a maximum of 10 regions in each sphere.

The modified subroutine 'Source' is shown in Table 8.13. As in Table 8.25P, there are three entry points in this routine. The first, entered by the call to 'Source' in subroutine 'Walk,' defines the source distribution for the first generation. We use the uniform source distribution over both spheres which is given in the version of 'Source' shown in Table 8.3. The second entry point, 'Sourc2,' is used to retrieve source particles from the fission bank in subsequent batches. This part of the current version of 'Source' is identical to that of Table 8.25P. The third entry point, 'Sourc3,' is used to define the matrix region in which each banked fission neutron is located. The present version tests the x coordinate of each fission event to determine the correct value of 'nreg' for each source particle.

The modified subroutine 'Col' is shown in Table 8.14. Functionally this routine is identical to that of Table 8.26P, but it uses appropriate changes in the matrix region definition to determine the value of the variable 'index.' The modified subroutine 'Stats' is shown in Table 8.15. The additions to entry point 'StatOne' made for the standard matrix calculation above are retained. The modifications at entry points 'StatEBatch' and 'StatEnd' are identical to those of Table 8.27P.

Table 8.12. Subroutine 'Input' Modified for Matrix by Generations, Exercise 8.1

```
      SUBROUTINE INPUT                                                          !   1
      COMMON/IN/npart,nbatch,nregions,iter
      WRITE(*,10)
   10 FORMAT(' Enter the number of regions in each sph. (max of 10) ')
      READ(*,*)nregions
      IF(nregions.LE.0.OR.nregions.GT.10)nregions=10
      WRITE(16,*)'Number of regions/sphere = ',nregions
      nregions = 2*nregions      !  double regions for two spheres
      WRITE(*,11)
   11 FORMAT(' Enter the number of matrix iterations. ')
      READ(*,*)iter
      WRITE(16,*)' Number of matrix iterations = ',iter
      WRITE(*,12)                                                               !   3
   12 FORMAT(' Enter the number of particles/batch.')
      READ(*,*)npart                                                            !   5
      WRITE(16,*)'Number of particles/batch = ',npart
      WRITE(*,14)
   14 FORMAT(' Enter the number of batches.')
      READ(*,*)nbatch
      WRITE(16,*)'Number of batches ',nbatch
      WRITE(*,13)                                                               !   7
   13 FORMAT(' Enter the random number seed, an integer. ')                     !   8
      READ(*,*)iseed                                                            !   9
      IF(iseed.LT.1)iseed=1                                                     !  10
      IF(iseed.GE.2147483647)iseed=2147483646                                   !  11
      WRITE(16,*)' Starting random number seed is ',iseed                       !  12
      CALL rndin(iseed)                                                         !  13
      RETURN                                                                    !  15
      END                                                                       !  16
```

Table 8.13. Subroutine 'Source' for Matrix by Generations, Exercise 8.1

```
      SUBROUTINE SOURCE                                                         !   1
      REAL(8) FLTRN
      COMMON/IN/npart,nbatch,nregions,iter
      REAL(8) x,y,z,u,v,w,xo,yo,zo,uo,vo,wo,wate,age,energ                      !   2
      COMMON/PART/x,y,z,u,v,w,xo,yo,zo,uo,vo,wo,wate,age,energ,nzcur,newzn,ngroup   !   3
      REAL(8) sphcx(20),sphcy(20),sphcz(20),sphr(20),rpxmin(20),rpxmax(20),&
       rpymin(20),rpymax(20),rpzmin(20),rpzmax(20),bdin(40),bdout(40)
      LOGICAL bdtest(40), bdhit(40)
      COMMON/cg/sphcx,sphcy,sphcz,sphr,rpxmin,rpxmax,rpymin,rpymax,rpzmin,rpzmax,&
       bdin,bdout,nsph,nrpp,nbdy,nzones,nbz(40,41),nnext(40,40,40,2),bdtest,bdhit
      REAL(8) sigt(20),c(20)
      COMMON/GEOM/sigt,c  ! sigt is total cross section, c is non-absorption probability
      REAL(8) xbank(200000,2),ybank(200000,2),zbank(200000,2),wbank(200000)
      COMMON/BANK/xbank,ybank,zbank,wbank,nsave
      REAL(8) bscore,bsumsq,cscore(2),csumsq(2),aij(20,20),aijsq(20,20),&
       bij(20,20),bijsq(20,20)
      COMMON/STAT/bscore,bsumsq,cscore,csumsq,aij,aijsq,&
       bij,bijsq,nstart(20),nbstart(20)
      COMMON/EX81/nreg,deltax,nregx
      pi=2.0d0*DACOS(0.0d0)          !  keep for more complicated source that needs pi   4
!  start 1st generation with uniform source distribution - fixed radius of 5
!  uniform distribution over both spheres
       DO 100 i=1,npart
         radius = 2.5*fltrn()**0.333333333
         costh = -1.+2.*fltrn()
         sinth = SQRT(1.-costh**2)
         phi = 2.*pi*fltrn()
```

Table 8.13. con't...

```
        zbank(i,2) = radius*costh
        xbank(i,2) = 3.+radius*sinth*COS(phi)
        ybank(i,2) = radius*sinth*SIN(phi)
        IF(fltrn().lt.0.5) xbank(i,2) = -xbank(i,2)
        wbank(i)=1.0d0
  100 CONTINUE
      nzcur = 1
      IF(x.lt.0.) nzcur=2
      nsave=npart  ;  RETURN
      ENTRY SOURC2
      initial=npart;  npart=0
      DO 200 i=1,nsave
  150 CONTINUE
      IF(wbank(i).GE.1.0d0)THEN
        npart=npart+1
        xbank(npart,1)=xbank(i,2);ybank(npart,1)=ybank(i,2);zbank(npart,1)=zbank(i,2)
        wbank(i)=wbank(i)-1.0d0  ;  GOTO 150
      ELSE
        IF(wbank(i).GE.FLTRN())THEN
          npart=npart+1
          xbank(npart,1)=xbank(i,2);ybank(npart,1)=ybank(i,2);zbank(npart,1)=zbank(i,2)
        ENDIF
      ENDIF
  200 CONTINUE
      IF(npart.GT.initial)npart=initial      !  keeps particle numbers from increasing
      nsave=0
      IF (npart.LE.0) THEN
        WRITE(*,*)' The number of particles decreased to zero.';  STOP
      ENDIF
      IF(npart.LT.initial)THEN               !  keeps particle numbers from decreasing
        ntemp=initial-npart
        DO i=1,ntemp
         npart=npart+1
         xbank(npart,1)=xbank(i,1);ybank(npart,1)=ybank(i,1);zbank(npart,1)=zbank(i,1)
        ENDDO
      ENDIF
      RETURN
      ENTRY SOURC3(n)
      x=xbank(n,1);  y=ybank(n,1);  z=zbank(n,1)
!  find matrix region for this source particle
      IF(x.LT.0.0) THEN      !  test sphere 1, x < 0
        xtest = -5.5
        DO 3 ij = 1,nregx
          xtest = xtest+deltax
          nreg = ij
          IF(x.le.xtest) GOTO 2
 3      CONTINUE
      ELSE          !  test sphere 2, x > 0
        xtest = 0.5
        DO 4 ij = 1, nregx
          nreg = nregx+ij
          xtest = xtest+deltax
          IF(x.le.xtest)GOTO 2
 4      CONTINUE
      ENDIF
 2    CONTINUE
      nstart(nreg)=nstart(nreg)+1;  nbstart(nreg)=nbstart(nreg)+1
      nzcur=1;  wate=1.0d0
      IF(x.lt.0.) nzcur = 2
      CALL ISOOUT                    !  direction chosen isotropically               7
      RETURN  ;  END
```

Table 8.14. Subroutine 'Col' for Matrix by Generations, Exercise 8.1

```
      SUBROUTINE COL                                                    !      1
      REAL(8) FLTRN,delta                                               !      2
      REAL(8) x,y,z,u,v,w,xo,yo,zo,uo,vo,wo,wate,age,energ              !      3
      COMMON/PART/x,y,z,u,v,w,xo,yo,zo,uo,vo,wo,wate,age,energ,nzcur,newzn,ngroup   ! 4
      REAL(8) dmfp,dtr,xsec,dcur                                        !      5
      COMMON/TRACK/dmfp,dtr,xsec,dcur                                   !      6
      REAL(8) sigt(20),c(20);  !  dimensions allow up to 20 different media     7
      COMMON/GEOM/sigt,c; !sigt is total cross section, c is non-absorption prob  8
      REAL(8) bscore,bsumsq,cscore(2),csumsq(2),aij(20,20),aijsq(20,20),&
       bij(20,20),bijsq(20,20)
      COMMON/STAT/bscore,bsumsq,cscore,csumsq,aij,aijsq,&
       bij,bijsq,nstart(20),nbstart(20)
      COMMON/IN/npart,nbatch,nregions,iter
      REAL(8) radius
      REAL(8) xbank(200000,2),ybank(200000,2),zbank(200000,2),wbank(200000)
      COMMON/BANK/xbank,ybank,zbank,wbank,nsave
      COMMON/EX81/nreg,deltax,nregx
      delta=dmfp/xsec                  !  distance traveled to collision        9
      dtr=dcur+delta                   !  update total distance traveled       10
      x=xo+u*dtr;  y=yo+v*dtr;  z=zo+w*dtr       !  update position            11
!  find index
       IF(x.LE.0.) THEN        !  sph 1, x<0
         index = 0
         DO 2 i = 1,nregx
           index = index+1
           IF(x.LT.FLOAT(i)*deltax-5.5) GOTO 3
 2       CONTINUE
         WRITE(*,*)i,nregx,index,deltax,x
         STOP 1
       ELSE            !  sphere 2, x>0
         index = nregx
         DO 4 i = 1, nregx
           index = index+1
           IF(x.lt.FLOAT(i)*deltax+0.5) GOTO 3
 4       CONTINUE
         WRITE(*,*)i,nregx,index,deltax,x
         STOP 2
       ENDIF
  3    CONTINUE
       aij(nreg,index)=aij(nreg,index)+c(nzcur)   !  scores all outgoing neutrons
       aijsq(nreg,index)=aijsq(nreg,index)+c(nzcur)**2
       bij(nreg,index)=bij(nreg,index)+c(nzcur)   !  scores all outgoing neutrons
       bijsq(nreg,index)=bijsq(nreg,index)+c(nzcur)**2
       nsave=nsave+1                    !  bank a fission
       xbank(nsave,2)=x                 !  x-position for fission
       ybank(nsave,2)=y                 !  y-position for fission
       zbank(nsave,2)=z                 !  z-position for fission
       wbank(nsave)=c(nzcur)            !  weight of fission particle
       cscore(1)=cscore(1)+c(nzcur)
       csumsq(1)=csumsq(1)+c(nzcur)**2
       nzcur=-1                         !  all incoming neutrons are absorbed
       RETURN;  END
```

Table 8.15. Subroutine 'Stats' for Matrix by Generations, Exercise 8.1

```
SUBROUTINE Stats                                                        !      1
       REAL(8) bscore,bsumsq,cscore(2),csumsq(2),aij(20,20),aijsq(20,20),&
        bij(20,20),bijsq(20,20)
       COMMON/STAT/bscore,bsumsq,cscore,csumsq,aij,aijsq,&
        bij,bijsq,nstart(20),nbstart(20)
       COMMON/IN/npart,nbatch,nregions,iter
       REAL(8) xbank(200000,2),ybank(200000,2),zbank(200000,2),wbank(200000)
       COMMON/BANK/xbank,ybank,zbank,wbank,nsave
       REAL(8) tmp,tmp1,tmp2,var,stdev,xkeff
```

Table 8.15. con't...

```
      REAL(8) sphcx(20),sphcy(20),sphcz(20),sphr(20),rpxmin(20),rpxmax(20),&
       rpymin(20),rpymax(20),rpzmin(20),rpzmax(20),bdin(40),bdout(40)
      LOGICAL bdtest(40), bdhit(40)
      COMMON/cg/sphcx,sphcy,sphcz,sphr,rpxmin,rpxmax,rpymin,rpymax,rpzmin,rpzmax,&
       bdin,bdout,nsph,nrpp,nbdy,nzones,nbz(40,41),nnext(40,40,40,2),bdtest,bdhit
      COMMON/EX81/nreg,deltax,nregx
ENTRY StatOne                 !  entry point to initialize arrays for complete problem  8
!  define nregion matrix regions as slices through the spheres
!  in Y-X plane but assume sphere centered at origin
      nregx = nregions/2
      deltax = 5./FLOAT(nregx)
!  write(*,*)nregx, nregions, deltax
      bscore=0.0d0;  bsumsq=0.0d0;  cscore=0.0d0;  csumsq=0.0d0;  outrad=sphr(1)
      aij=0.0d0;  aijsq=0.0d0;  nstart=0;  bij=0.0d0;  bijsq=0.0d0;  nbstart=0
RETURN                                                                          !   10
ENTRY StatLp                  !  entry point to initialize arrays for a particle     11
RETURN                                                                          !   13
ENTRY StatELp                 !  entry point to store scores for a particle          14
RETURN                                                                          !   19
ENTRY StatEBatch                   !  entry point to store scores after each batch
      tmp1=DFLOAT(npart);  tmp2=0.d0
      DO 100 i=1,nsave
      tmp2=tmp2+wbank(i)
  100 CONTINUE
      tmp=tmp2/tmp1;  bscore=bscore+tmp;  bsumsq=bsumsq+tmp**2
      DO i=1,nregions
        tmp=DFLOAT(nbstart(i))
        DO j=1,nregions
          var=bijsq(i,j)/tmp-(bij(i,j)/tmp)**2
          bij(i,j)=bij(i,j)/tmp
          bijsq(i,j)=var/tmp
        END DO
      END DO
      CALL keff(bij,bijsq,xkeff,var)
      cscore(2)=cscore(2)+xkeff;  csumsq(2)=csumsq(2)+xkeff**2
      bij=0.0d0;  bijsq=0.0d0;  nbstart=0
RETURN
ENTRY StatEnd                 !  entry point to calculate and print results            20
      tmp=DFLOAT(nbatch)
      var=bsumsq/tmp-(bscore/tmp)**2;  stdev=DSQRT(var)
      WRITE(16,*)'ratio of gen      ',bscore/tmp,stdev/DSQRT(tmp)
      tmp=DFLOAT(nbatch)*DFLOAT(npart)
      var=csumsq(1)/tmp-(cscore(1)/tmp)**2;  stdev=DSQRT(var)
      WRITE(16,*)'total subseq pop ',cscore(1)/tmp,stdev/DSQRT(tmp)
      tmp=DFLOAT(nbatch)
      var=csumsq(2)/tmp-(cscore(2)/tmp)**2;  stdev=DSQRT(var)
      WRITE(16,*)'matrix by gen    ',cscore(2)/tmp,stdev/DSQRT(tmp)
      DO i=1,nregions
        tmp=DFLOAT(nstart(i))
        DO j=1,nregions
          var=aijsq(i,j)/tmp-(aij(i,j)/tmp)**2
          aij(i,j)=aij(i,j)/tmp
          aijsq(i,j)=var/tmp
        END DO
      END DO
      CALL keff(aij,aijsq,xkeff,var)
      WRITE(16,200)xkeff,DSQRT(var)
  200 FORMAT(' standard matrix   k-eff = ',1pe17.8,' +- ',1pe17.8)
      WRITE(16,*)(nstart(i),i=1,nregions)
RETURN                                                                          !   28
      END                                                                       !   29
```

The results obtained by running this modified code with the same six regions shown in Table 8.10, using c = 1.27435, three regions per sphere, 25 matrix iterations, 10^4 source particles per batch, and 25 batches, are shown in Table 8.16. As expected, we obtain a value of k quite close to unity even when using this small number of regions. The results for c = 1.27435, 20 regions, 25 matrix iterations, 10^5 start particles per batch, and 25 batches, also shown in Table 8.16, indicate that this combined generation and matrix calculation indeed allows the selection of coarse matrix regions without sacrificing accuracy. The fission event distributions by region obtained in these calculations are shown in Table 8.17. By symmetry, for m total regions the number of fission events in region n should be equal to the number of fissions in region m-n+1. We see that this is approximately the case, which supports the conclusion that we have obtained a reasonable result.

Table 8.16. Matrix-by-generation Results for Exercise 8.1

	6 Regions	20 Regions
Ratio of Generations	0.99927 ± 0.00122	0.99948 ± 0.00028
Total Subsequent Population	0.99927 ± 0.00105	0.99948 ± 0.00033
Matrix by Generation	1.00000 ± 0.00121	0.99953 ± 0.00028
Standard Matrix	0.99934 ± 0.00175	0.99942 ± 0.00076

Table 8.17. Fission Events by Region, Matrix-by-Generation Results for Exercise 8.1

m = 6, n=(1,6)						
	26709	69589	33083	31225	64758	24636
m = 20, n=(1,20)						
	18024	67540	124483	177408	208048	214831
192636	145830	87976	27302	27366	87737	143322
187354	210090	205282	171089	121548	64757	17377

Exercise 8.2

Statement of the problem

2. Find the critical radius and critical mass of fissile material for a homogeneous sphere of ^{235}U and water. Assume isotropic scatter in the center of mass to account for neutron downscatter in the water and use the Watt fission spectrum for fission neutrons. Assume a uranium density of 1.5 g/cm^3. Use the Hansen-Roach 16-group cross sections for ^{235}U. Use the subroutines 'Hydrogen' and 'Oxygen' from Table 4.10P for water. [Hint: Assume mid-energy of each group for all uranium interactions, continuous energy in water.]

Solution

We assume for this calculation that we have, in effect, a uranium salt uniformly dissolved in water. We assume that the stated uranium density of 1.5 g/cm^3 gives an effective volume fraction of uranium that is equal to this density divided by the full density of uranium. The latter is given for the Godiva assembly on p. 259 of the *Primer* as 18.80 g/cm^3. Using this value as the full density of uranium, the volume fraction of uranium in the postulated mixture is 0.07979. Assuming the remaining material is water, this result in turn gives the volume fraction of water in the mixture as 0.9202. From these volume fractions we obtain the following atom densities in atoms/barn-cm: $N_U = 0.003844$, $N_H = 0.06158$, $N_O = 0.03079$.

The problem statement does not define the method we should use to solve the exercise. Therefore we are free to choose any of the methods described in the *Primer*. Rather than walk the reader through each option as we did in Exercise 8.1 we use the generation and total-subsequent-population methods to solve the problem. Following the procedure developed for Example 8.1 we modify the PFC subroutines listed in Table 8.18.

Table 8.18. Modified Subroutines for Exercise 8.2

Subroutine	Location
'Hydrogen'	Table 4.10P
'Oxygen'	Table 4.10P
'Input'	Table 8.7P
'Walk'	Table 8.19
'Stats'	Table 8.20
'Source'	Table 8.21
'Col'	Table 8.22
'Isocol'	Table 6.22
'Xsects'	Table 8.23
'Mxsec'	Table 8.24

The modified subroutine 'Walk' is shown in Table 8.19. The only difference between this routine and that in Table 8.3P is the value of 'ntransient.' For this exercise, as is discussed below, we use 'ntransient' = 5. The modified subroutine 'Stats' is shown in Table 8.20. This routine is changed to enable the user to write the batch k on the output file as is done in Example 8.1, but is otherwise identical to the version of Table 8.11P. The write statements are commented out in the table.

The modified subroutine 'Source' is shown in Table 8.21. This subroutine is similar to that of Table 8.36P but includes a number of changes. For the present calculation, involving a homogeneous, unreflected sphere of material, we use a uniform initial source distribution over the sphere. This spatial distribution is defined in the 'do' loop number 100 of subroutine 'Source.' Because we wish to combine the Hansen-Roach 16-

group cross sections for ^{235}U, shown in Table 8.31P, with the continuous-slowing-down model developed in Chapter 4 of the *Primer*, we need to keep a record of the energy of the neutrons as they experience collisions with the three elements in this problem: H, O, and U. We therefore add the array 'ebound,' which contains the lower energy bounds of the Hansen-Roach groups taken from Table 8.30P, to subroutine 'Source.'

Table 8.19. Subroutine 'Walk' Modified for Exercise 8.2

```
      SUBROUTINE WALK                                                    !    1
      COMMON/IN/npart,nbatch                                             !    2
      REAL(8) dmfp,dtr,xsec,dcur                                         !    3
      COMMON/TRACK/dmfp,dtr,xsec,dcur                                    !    4
      REAL(8) x,y,z,u,v,w,xo,yo,zo,uo,vo,wo,wate,age,energ               !    5
      COMMON/PART/x,y,z,u,v,w,xo,yo,zo,uo,vo,wo,wate,age,energ,nzcur,newzn,ngroup  !  6
      ntransient=5
      CALL SOURCE                  ! get source parameters for initial bank
Loop_Over_Batches:  DO j=1,nbatch+ntransient
      IF(j.EQ.ntransient+1)CALL StatOne          ! re-initializes statistics arrays
      CALL SOURC2
Loop_Over_Particles: DO i=1,npart   ! loop over number of particles          7
      CALL StatLp                  ! initialize statistics for each particle  8
      CALL SOURC3(i)
 Loop_for_Collisions: DO           ! to find collision sites                 10
      CALL DIST                    ! get distance of travel in mfp           11
    Loop_Track:  DO                ! track particle to collision point       12
      CALL HIT                     ! get distance to boundary of current zone 13
      CALL MXSEC                   ! get total cross section for current zone 14
      IF((dtr-dcur)*xsec.GE.dmfp) EXIT Loop_Track  ! collision before boundary 15
      CALL BDRX                    ! process boundary crossing               16
      IF(nzcur.EQ.-1) EXIT Loop_for_Collisions  ! particle killed e.g., by escaping 17
    END DO Loop_Track                                               !        18
      CALL COL                     ! call col to process collision           19
      IF(nzcur.EQ.-1) EXIT Loop_for_Collisions  ! particle killed by collision 20
  END DO Loop_for_Collisions                                        !        21
      CALL StatELp                 ! process statistics when particle killed 22
END DO Loop_Over_Particles                                          !        23
      IF(j.GT.ntransient)CALL StatEBatch  ! process statistics when particle killed
END DO Loop_Over_Batches
      RETURN                                                        !        24
      END                                                           !        25
```

Table 8.20. Subroutine 'Stats' Modified for Exercise 8.2

```
  SUBROUTINE Stats                                                       !    1
      REAL(8) bscore,bsumsq,cscore,csumsq,cpart
      COMMON/STAT/bscore,bsumsq,cscore,csumsq,cpart
      COMMON/IN/npart,nbatch                                             !    4
      REAL(8) x,y,z,u,v,w,xo,yo,zo,uo,vo,wo,wate,age,energ               !    5
      COMMON/PART/x,y,z,u,v,w,xo,yo,zo,uo,vo,wo,wate,age,energ,nzcur,newzn,ngroup  !  6
      REAL(8) tmp,tmp1,tmp2,var,stdev                                    !    7
      REAL(8) xbank(20000,2),ybank(20000,2),zbank(20000,2),wbank(20000)
      COMMON/BANK/xbank,ybank,zbank,wbank,nsave
  ENTRY StatOne          ! entry point to initialize arrays for complete problem  8
      bscore=0.0d0;  bsumsq=0.0d0;  cscore=0.0d0;  csumsq=0.0d0
      WRITE(16,1);  nbchpr = 0
1     FORMAT(/,' Batch k')
  RETURN                                                                 !   10
  ENTRY StatLp           ! entry point to initialize arrays for a particle       11
      cpart=0.0d0
```

Table 8.20. con't...

```
RETURN                                                               !     13
ENTRY StatELp                  ! entry point to store scores for a particle  14
      cscore=cscore+cpart
      csumsq=csumsq+cpart**2
RETURN                                                               !     19
ENTRY StatEBatch
      tmp1=DFLOAT(npart);  tmp2=0.d0
      DO 100 i=1,nsave
        tmp2=tmp2+wbank(i)
 100 CONTINUE
      tmp=tmp2/tmp1;  bscore=bscore+tmp;  bsumsq=bsumsq+tmp**2
      nbchpr = nbchpr+1
!      WRITE(*,*)  nbchpr, tmp
!      WRITE(16,*) nbchpr, tmp
RETURN
ENTRY StatEnd                  ! entry point to calculate and print results   20
      WRITE(16,2)
2     FORMAT(/,' keff estimates from gen k and tot sub pop k')
      tmp=DFLOAT(nbatch)
      var=bsumsq/tmp-(bscore/tmp)**2           ! variance of bscore distribution
      stdev=DSQRT(var)                         ! standard deviation of bscore distr23
      WRITE(16,*)bscore/tmp,stdev/DSQRT(tmp)   ! result and std dev of result
      tmp=DFLOAT(nbatch)*DFLOAT(npart)
      var=csumsq/tmp-(cscore/tmp)**2
      stdev=DSQRT(var)  .
      WRITE(16,*)cscore/tmp,stdev/DSQRT(tmp)
RETURN                                                               !     28
      END                                                            !     29
```

To find the average energy for each of the Hansen-Roach groups one should properly use flux weighting; however, for present purposes the average energy is estimated on the basis of the average group velocity given in Table 8.30P. We use eqn 7.44P to obtain the average energies from the velocities, and these average energies are placed in the array 'ebar.' The resulting group-average energies are considered reasonable for this exercise. Both the group energy bounds and the group average energies are specified in data statements in subroutine 'Source.' They are passed to the other routines through common 'e82.' The variable 'energ' in common 'part' is used to define the discrete energy of a neutron at any point in its path, while the variable 'ngroup' is used to define the Hansen-Roach energy group for the particle of energy 'energ.'

The coding under the entry point 'Sourc2' is unchanged from that of Table 8.36P. This coding retrieves the location and weight of each fission neutron in a batch based on the fissions in the previous batch. As in Example 8.2 each fission event is assumed to result in the capture of the incident neutron, and all fission neutrons are assumed to appear in the next generation. See the discussion on p. 232 of the *Primer* for the importance of this parameter when the multiplication of the system is not equal to one.

Entry point 'Sourc3' is modified to use the Watt fission spectrum to determine the energy of the fission neutrons, as specified in the problem definition. The coding for the Watt spectrum is taken from that in Table

4.7P. After the energy of the new fission neutron is set, the Hansen-Roach energy group for this neutron is determined so that it will be available for use should the neutron collide with a ^{235}U nucleus.

Table 8.21. Subroutine 'Source' Modified for Exercise 8.2

```
      SUBROUTINE SOURCE                                                     !     1
      REAL(8) FLTRN,tmp
      COMMON/IN/npart,nbatch
      REAL(8) x,y,z,u,v,w,xo,yo,zo,uo,vo,wo,wate,age,energ                  !     2
      COMMON/PART/x,y,z,u,v,w,xo,yo,zo,uo,vo,wo,wate,age,energ,nzcur,newzn,ngroup  !  3
      REAL(8) sphcx(20),sphcy(20),sphcz(20),sphr(20),rpxmin(20),rpxmax(20),&
       rpymin(20),rpymax(20),rpzmin(20),rpzmax(20),bdin(40),bdout(40)
      LOGICAL bdtest(40), bdhit(40)
      COMMON/cg/sphcx,sphcy,sphcz,sphr,rpxmin,rpxmax,rpymin,rpymax,rpzmin,rpzmax,&
       bdin,bdout,nsph,nrpp,nbdy,nzones,nbz(40,41),nnext(40,40,40,2),bdtest,bdhit
      REAL(8) xbank(20000,2),ybank(20000,2),zbank(20000,2),wbank(20000)
      COMMON/BANK/xbank,ybank,zbank,wbank,nsave
      REAL(8) sigt(16),c(16),pscat(16),pabs(16),&
       pfis(16),gpnu(16),pggscat(16,16)
      COMMON/GEOM/sigt,c,pscat,pabs,pfis,gpnu,pggscat,vel,chi,nerggp
!  energ is the current neutron energy in eV
!  ebound are the lower energy bounds of the H-R 16-group set
!  ebar are the group average energies
      REAL(8) hxs,oxs,uxs
      COMMON/e82/ebound(16),ebar(16),hxs,oxs,uxs
      DATA ebound/3.e6, 1.4e6, 0.9e6, 0.4e6, 0.1e6, 17000., 3000.,&
       550., 100., 30., 10., 3., 1., 0.4, 0.1, 0.0/
      DATA ebar/4.25e+6, 2.07e+6, 1.13e+6, 0.633e+6, 0.235e+6, &
       38.1e+3, 6.79e+3, 1.2e+3, 222., 53.3, 16.7, 5.32, 1.68, &
       0.621, 0.192, 0.025/
!  uniform source assumed
      DO 100 i=1,npart
        wbank(i)=1.0d0
        r=sphr(1)*FLTRN()**0.333333333      !  use for uniform source
        costh = -1.+2.*fltrn()
        stheta = SQRT(1.-costh**2)
        phi = 6.2831853*fltrn()
        xbank(i,2) = r*stheta*COS(phi)
        ybank(i,2) = r*stheta*SIN(phi)
        zbank(i,2) = r*costh
  100 CONTINUE
      nsave=npart
      RETURN
      ENTRY SOURC2
      initial=npart;  npart=0
      DO 200 i=1,nsave
  150 CONTINUE
      IF(wbank(i).GE.1.0d0)THEN
        npart=npart+1
        xbank(npart,1)=xbank(i,2); ybank(npart,1)=ybank(i,2); zbank(npart,1)=zbank(i,2)
        wbank(i)=wbank(i)-1.0d0
        GOTO 150
      ELSE
        IF(wbank(i).GE.FLTRN())THEN
          npart=npart+1
          xbank(npart,1)=xbank(i,2);ybank(npart,1)=ybank(i,2);zbank(npart,1)=zbank(i,2)
        ENDIF
      ENDIF
  200 CONTINUE
      IF(npart.GT.initial)npart=initial      !  keeps particle numbers from increasing
      nsave=0
```

Table 8.21. con't...

```
      IF (npart.LE.0) THEN
        write(*,*)' The number of particles decreased to zero.'
        STOP
      ENDIF
      IF(npart.LT.initial)THEN                   ! keeps particle numbers from decreasing
        ntemp=initial-npart
        DO i=1,ntemp
          npart=npart+1
          xbank(npart,1)=xbank(i,1);ybank(npart,1)=ybank(i,1);zbank(npart,1)=zbank(i,1)
        ENDDO
      ENDIF
      RETURN
      ENTRY SOURC3(n)
      x=xbank(n,1);  y=ybank(n,1);  z=zbank(n,1)
      nzcur=1;  wate=1.0d0
!  select energy from Watt fission spectrum
 65   xtest=-dlog(fltrn());ytest=-dlog(fltrn())        ! pick a point
      IF((ytest-xtest-1.)**2.gt.4.*xtest) GOTO 65      ! reject if fails test
      energ = 2.*xtest*1.e6                      ! value selected from fission spectrum
!  find group number for this energy
      DO 300 i=1,nerggp
        IF(energ.gt.ebound(i))GOTO 301
 300  CONTINUE
 301  ngroup=i
      CALL ISOOUT                  ! direction chosen isotropically                      7
      RETURN                                                                        !    9
      END                                                                           !   10
```

The modified subroutine 'Col' is shown in Table 8.22. This subroutine retains the structure of the routine shown in Table 8.37P but is modified to reflect the fact that two types of collisions are possible in this problem. Collisions with a ^{235}U nucleus are treated with the Hansen-Roach multigroup cross sections for that isotope, as shown in Table 8.31P, and the coding for such collisions in the present version of subroutine 'Col' is similar to that of Table 8.37P. However, collisions with either H or O require the use of the isotropic center-of-mass scattering kernel of Chapter 4 of the *Primer*. To determine which type of nucleus a neutron has hit we place the macroscopic cross sections for the three target nuclei, 'hxs,' 'oxs,' and 'uxs,' at the current neutron energy, in common 'e82.' These cross sections are obtained from subroutine 'Mxsec,' as explained below. The probability of the neutron encountering each of the three types of nuclei in the problem is equal to the ratio of the macroscopic cross section for that type of nucleus divided by the total macroscopic cross section, 'xsec,' for neutrons of the current energy. That is, since

$$\Sigma_t = \Sigma_H + \Sigma_O + \Sigma_U$$

we have

$$P_i = \frac{\Sigma_i}{\Sigma_t}$$

where P_i is the probability of the neutron striking a nucleus of material i, i = H, O, or ^{235}U.

Table 8.22. Subroutine 'Col' Modified for Exercise 8.2

```
      SUBROUTINE COL                                                      !     1
      REAL(8) FLTRN,delta,r,a
      REAL(8) x,y,z,u,v,w,xo,yo,zo,uo,vo,wo,wate,age,energ                !     3
      COMMON/PART/x,y,z,u,v,w,xo,yo,zo,uo,vo,wo,wate,age,energ,nzcur,newzn,ngroup  ! 4
      REAL(8) dmfp,dtr,xsec,dcur                                          !     5
      COMMON/TRACK/dmfp,dtr,xsec,dcur                                     !     6
      REAL(8) sigt(16),c(16),pscat(16),pabs(16),&
       pfis(16),gpnu(16),pggscat(16,16)
      COMMON/GEOM/sigt,c,pscat,pabs,pfis,gpnu,pggscat,vel,chi,nerggp
      REAL(8) xbank(20000,2),ybank(20000,2),zbank(20000,2),wbank(20000)
      COMMON/BANK/xbank,ybank,zbank,wbank,nsave
      REAL(8) bscore,bsumsq,cscore,csumsq,cpart
      COMMON/STAT/bscore,bsumsq,cscore,csumsq,cpart
      REAL(8) hxs,oxs,uxs
      COMMON/e82/ebound(16),ebar(16),hxs,oxs,uxs
      delta=dmfp/xsec                    ! distance traveled to collision       9
      dtr=dcur+delta                     ! update total distance traveled      10
      x=xo+u*dtr;  y=yo+v*dtr;  z=zo+w*dtr      ! update position              11
!  determine which type of nucleus was struck
      r = fltrn()
      IF(r.LT.(hxs+oxs)/xsec)THEN         ! hits either O or H
        a = 16.                           ! might be O
        IF(r.LT.hxs/xsec)THEN             ! hits H
          a = 1.
        ENDIF
        CALL isocol(u,v,w,a,energ)
        IF(energ.lt.ebar(16)) energ = ebar(16)
        DO 4 i = 1,nerggp
          ngroup = i
          IF(energ.gt.ebar(i)) EXIT
 4      CONTINUE
        RETURN
      ENDIF
!  hits U235
      IF(FLTRN().LE.pfis(ngroup))THEN      ! if fission
        nsave=nsave+1                     ! bank a fission
        xbank(nsave,2)=x                  ! x-position for fission
        ybank(nsave,2)=y                  ! y-position for fission
        zbank(nsave,2)=z                  ! z-position for fission
        wbank(nsave)=wate*gpnu(ngroup)      ! weight of fission particle
        cpart=cpart+wate*gpnu(ngroup)
      ENDIF
      wate=wate*pscat(ngroup)      ! reduce wate by non-absorption
      IF(wate.LT.0.1d0)THEN               ! russian roulette to kill if wate is small
          IF(wate.LT.FLTRN())THEN         ! killed by RR
             nzcur=-1;  RETURN
          ENDIF
          wate=1.0d0                      ! survived, set weight = 1
      ENDIF
!     find new energy group
      r=FLTRN()
      DO 100 i=1,nerggp
        IF(r.LT.pggscat(ngroup,i))GOTO 101
 100 CONTINUE
 101 ngpold = ngroup ;  ngroup=i
      IF(ngroup.NE.ngpold) energ = ebar(ngroup)  ! no change for within-group scatter
      CALL ISOOUT                         ! assumes isotropic scatter in lab system   20
      RETURN                                                              !    21
      END                                                                 !    22
```

In this calculation a neutron is always assigned both a discrete energy, 'energ,' and a Hansen-Roach energy group, 'ngroup.' The energy bounds of the latter bracket the energy 'energ.' Because of this dual energy tracking, when the neutron hits a uranium atom we determine the result of the collision in a different manner than is done in Table 8.37P. To assign a discrete energy to neutrons after they experience a collision with a ^{235}U atom, we treat within-group scattering, which involves no energy loss in the multigroup format, different from group-to-group scattering, which involves energy downscatter. When the neutron leaves a collision with uranium in a lower energy group than the group it was in when it entered the collision we assign to the post-collision neutron the average energy of the Hansen-Roach group into which it has scattered. When a collision with a ^{235}U atom results in the post-collision neutron remaining in the same energy group we assume the energy of the neutron is unchanged by the collision.

When continuous and multi-group energy treatments are combined in the same calculation the discrete post-collision neutron energy following a multi-group scattering event can be defined in several ways. For example, for within-group scattering the post-collision neutron energy could be set to the group-average energy. For neutron downscatter the post-collision energy could be selected from some assumed energy distribution in the new group. The multi-group cross sections provide no information about the flux shape within each group, but in all cases some flux distribution was used to obtain the group-average data. Ideally this flux distribution is the same as that in the calculation and could be used to select a discrete neutron energy.

If post-collision neutrons are assigned the group-average energy following within-group scattering, some incident neutrons will lose energy while others will gain energy. One might argue that this is correct because the multi-group cross section already reflects the flux distribution in the group. In narrow energy groups such treatment is probably acceptable; however, in this problem the energy groups are broad and neutron downscatter to a lower energy group following a collision with ^{235}U is unlikely. Most of the neutron energy loss results from scattering on H and O. In the latter events multi-group cross sections are not used. As we have seen in Example 4.1, even with H as the scattering target several scattering events are necessary on the average to reduce the neutron energy significantly. With the Hansen-Roach group structure setting the post-collision neutron energy to the group average following within-group scattering could artificially tie the neutrons to the incident group; i.e., unless several collisions with H or O occur between successive ^{235}U collisions, raising the neutron energy to the group average value may cause the neutron to stay in the same energy group much longer than is physically correct. On the other hand, retaining the incident energy in within-group scattering events allows the neutrons to scatter out of broad energy groups following a sufficient number of collisions with H and O.

Assigning a post-collision neutron the average energy in a downscatter group following a scattering event with ^{235}U does not carry the same implication as doing so following within-group scattering. In the downscatter case neutrons that should physically have energies greater than the average value, but are assigned less, are compensated by those that should have less than the average but are assigned more. Because the multi-group cross sections provide no information about how the downscattered neutrons should be distributed within their new energy group we elect to assign the group average energy to all.

In the current calculation Russian roulette is played only after a collision with a ^{235}U atom. If a neutron survives Russian roulette after encountering a ^{235}U nucleus, the post-collision direction of travel of the neutron is defined isotropically by calling subroutine 'Isoout.' The coding could easily be changed to perform Russian roulette on neutrons after collisions with H or O if the exercise assumptions were changed.

In subroutine 'Col' if the neutron strikes a H or O atom the mass of the scatterer, 'a,' is defined and the center-of-mass scattering routine 'Isocol' is called. The version of subroutine 'Isocol' given in Table 6.22 is used here. Except for the use of double precision variables, this routine is identical to that of Table 4.9P. Once the post-collision neutron energy 'energ' following a collision with H or O has been defined, the coding in subroutine 'Col' determines the Hansen-Roach energy group for neutrons of that energy.

The modified subroutine 'Xsects' is shown in Table 8.23. Instead of using a special mixing routine, as is done in Example 8.4 and as shown in Table 8.33P, here the ^{235}U cross sections are mixed in subroutine 'Xsects.' Thus both the number of energy groups and the microscopic ^{235}U cross sections of Table 8.31P are read from file 'xsects.txt' directly into subroutine 'Xsects.' The file 'xsects.txt' is structured identically to that read by the program in Table 8.33P. That is, the number of energy groups is followed by the data of Table 8.31P, row by row, with the blanks being replaced by zeros in the scatter matrix and zeros being included for all possible downscatter groups, not just those shown in the table. The arrays 'pfis,' 'pabs,' 'sigt,' 'gpnu,' and 'pggscat' have the same meanings they did in Tables 8.33P and 8.38P, but here they are dimensioned for only a single element in the 16-group cross section set. The microscopic data are converted to cumulative probabilities in the 'DO' loop number 250 in the same manner as is done in Table 8.33P.

Table 8.23. Subroutine 'Xsects' Modified for Exercise 8.2

```
      SUBROUTINE XSECTS                                                   !    1
      REAL(8) sigt(16),c(16),pscat(16),pabs(16),&
       pfis(16),gpnu(16),pggscat(16,16),s
      COMMON/GEOM/sigt,c,pscat,pabs,pfis,gpnu,pggscat,vel,chi,nerggp
      REAL(8) sphcx(20),sphcy(20),sphcz(20),sphr(20),rpxmin(20),rpxmax(20),& !   4
       rpymin(20),rpymax(20),rpzmin(20),rpzmax(20),bdin(40),bdout(40)     !    5
      LOGICAL bdtest(40), bdhit(40)                                       !    6
      COMMON/cg/sphcx,sphcy,sphcz,sphr,rpxmin,rpxmax,rpymin,rpymax,rpzmin,rpzmax,& ! 7
       bdin,bdout,nsph,nrpp,nbdy,nzones,nbz(40,41),nnext(40,40,40,2),bdtest,bdhit ! 8
!  read microscopic cross sections and calculate cumulative probabilities
      OPEN(UNIT=12,FILE='xsects.txt')        !  The cross sections are in xsects.txt
      READ(12,*)nerggp        !  read number of energy groups
!  read U-235 cross sections: gp #, fission, absorp, tot, gp-to-gp scatt
      DO 1 i = 1,nerggp
 1    READ(12,*)n,pfis(i),pabs(i),sigt(i),gpnu(i),&
       (pggscat(i,j),j=i,nerggp)
!  calculate cumulative probabilities and
!  convert gpnu from nu*sigf to nu by dividing by sigf
     DO 250 i=1,nerggp
        gpnu(i)=gpnu(i)/pfis(i)               !  gpnu=nu*sigf/sigf
        pfis(i)=pfis(i)/sigt(i)               !  get prob of fission
        pabs(i)=pabs(i)/sigt(i)               !  get prob of absorption w/o fission
        pscat(i)=1.d0-pabs(i)-pfis(i)         !  all other prob is scatter
        s=0.0d0
        DO 230 j=1,nerggp
        s=s+pggscat(i,j)                      !  get normalization factor
 230    CONTINUE
        DO 240 j=1,nerggp
          pggscat(i,j)=pggscat(i,j)/s         !  normalize so sum is one
 240    CONTINUE
        pggscat(i,nerggp)=1.0d0
        DO 245 j=2,nerggp-1
          pggscat(i,j)=pggscat(i,j-1)+pggscat(i,j)      !  need cummulative prob
 245    CONTINUE
 250 CONTINUE
      RETURN                                                              !   16
      END                                                                 !   17
```

The modified subroutine 'Mxsec' is shown in Table 8.24. This routine determines the total cross section 'xsec' for a neutron at the current energy 'energ' for use in calculating the collision point for the current path. To account for the three species involved in this calculation, common 'e82' is included in the routine along with common 'part.' The latter includes the current energy 'energ' and the current group 'ngroup.' The number densities of the three species, given above, are included as a data statement in 'Mxsec' so that the macroscopic cross sections can be determined. The microscopic H and O cross sections are obtained from the routines 'Hydrogen' and 'Oxygen' as is done in Example 4.3. These routines, shown in Table 4.10P, return the cross section values in single precision. Although they could be modified to include double precision as in the routines of Tables 7.32P and 7.33P, we use them as written and convert the result to double precision in 'Mxsec.' The ^{235}U microscopic cross sections are stored in the array 'sigt' in subroutine 'Xsects,' and the macroscopic ^{235}U cross section at the current neutron energy is calculated in 'Mxsec' using this array. The total cross section is then calculated and passed to subroutine 'Walk.'

Table 8.24. Subroutine 'Mxsec' Modified for Exercise 8.2

```
      SUBROUTINE MXSEC                                                    !  1
      REAL(8) dmfp,dtr,xsec,dcur                                          !  2
      COMMON/TRACK/dmfp,dtr,xsec,dcur                                     !  3
      REAL(8) x,y,z,u,v,w,xo,yo,zo,uo,vo,wo,wate,age,energ                !  4
      COMMON/PART/x,y,z,u,v,w,xo,yo,zo,uo,vo,wo,wate,age,energ,nzcur,newzn,ngroup  !  5
      REAL(8) sigt(16),c(16),pscat(16),pabs(16),&
       pfis(16),gpnu(16),pggscat(16,16)
      COMMON/GEOM/sigt,c,pscat,pabs,pfis,gpnu,pggscat,vel,chi,nerggp
      REAL(8) hxs,oxs,uxs
      COMMON/e82/ebound(16),ebar(16),hxs,oxs,uxs
!  number densities based on vol frac of U
      DATA hdens/0.06158/, odens/0.03079/, udens/0.003844/
      energl = energ                       !  current neutron energy in single-precision
      CALL hydrogen(hxsl, energl, hdens)   !  get single-precision hydrogen cross section
      CALL oxygen(oxsl, energl, odens)     !  get single-precision oxygen cross section
      hxs=hxsl; oxs=oxsl         !  convert H and O cross sections to double precision
      uxs=sigt(ngroup)*udens  ! get U235 total cross section (no premixed cross sections)
      xsec = hxs+oxs+uxs      ! calculate total cross section
      RETURN                                                              !  9
      END                                                                 !  10
```

The geometry input for this calculation is shown in Table 8.25. One spherical body, centered at the origin, is used to define the problem geometry. The radius of the sphere, which is set to 15 cm in this table, is modified to determine the neutron multiplication of the system as a function of radius. Because the density of uranium in the system is known, the mass of uranium is determined from the radius.

The coding for subroutine 'Walk' in Table 8.19 sets the number of generations used to establish an equilibrium distribution of fissions in the system, defined by the variable 'ntransient,' to five. To confirm that this is a reasonable number we set 'ntransient' to zero and plot the resulting batch k values. Running PFC with the above modifications, which assumes a uniform source distribution for the first batch, using 5000 start particles per generation and a system radius of 15 cm, gives the results shown in Figure 8.4. The convergence to a normal mode occurs quickly. Therefore we conclude that setting 'ntransient' equal to five is reasonable.

Table 8.25. Contents of the File 'geom.txt' for Exercise 8.2

0	No RPPs
1	One SPH
0. 0. 0. 15.	Location and radius of SPH
2	Number of zones
1	No. bodies in zone 1
1	Definition of zone 1
1	No. bodies in zone 2
-1	Definition of zone 2

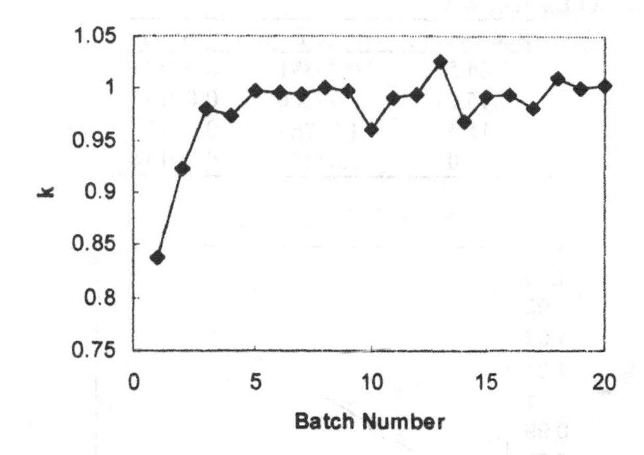

Figure 8.4. System Multiplication as a Function of Batch,

Running the modified PFC as given above for 10^4 start particles per batch for 100 batches, with 'ntransient' set to five, and for various system radii, gives the results shown in Table 8.26. The modifications given above provide both the standard generation k and the total-subsequent-population k estimates, which are identical, along with estimates of the standard deviations for both. The standard deviation estimated for the total-subsequent-population score provides the better of the two uncertainty estimates and therefore we show only this error estimate in the table. These results are plotted in Figure 8.5.

We use linear regression to obtain a least-squares fit to the data in Table 8.26. For these data we obtain the fit

$$k = 0.05337\, r + 0.19056$$

We do not expect the actual functional relationship between the radius and the effective neutron multiplication to be linear, but we can assume linearity over small ranges in the radius r. Alternatively we could use a quadratic fit to estimate k. Using the above expression for k we obtain a critical radius of about 15.166 cm, or a volume of 1.46×10^4 cm^3. At a uranium density of 1.5 g/cm^3 this gives a critical mass of approximately 21.9 kg of ^{235}U. We can confirm this result by running the code with the geometry set to this system radius. Performing this calculation with 100 batches at 10^4 particles per batch we obtain keff $\approx$ 1.00055 $\pm$ 0.00133. Thus the actual critical radius, based on the current model, is confirmed to be close to 15.166 cm.

Table 8.26. Results of Exercise 8.2

Radius (cm)	keff	sigma
14.5	0.96391	0.00131
15.0	0.99206	0.00132
15.5	1.01763	0.00133
16.0	1.04434	0.00134

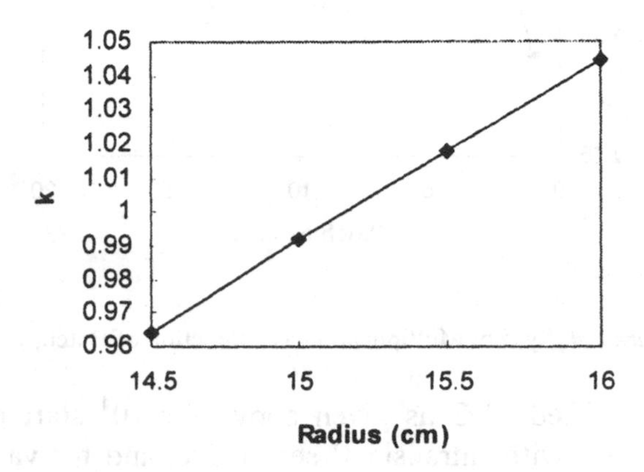

Figure 8.5. Results of Exercise 8.2

Exercise 8.3

Statement of the problem

3. Use the matrix method to calculate the effective neutron multiplication in a system containing ^{235}U spheres in water. Assume there are nine uranium spheres, each 3 cm in diameter, placed on a rectangular grid with 5 cm between the centers of the uranium spheres. The intervening space is filled with water and the outer boundary of the system is 2 cm outside the outermost uranium spheres. Assume pure ^{235}U with a density of 18.80 g/cm^3. Use the Exercise 8.2 cross section assumptions.

Solution – Standard Matrix k

Because the geometry for this exercise is neatly divided into nine equal parts this problem appears to offer an obvious choice for using these parts to define the regions for a matrix criticality calculation. Therefore we perform a matrix k calculation using each of the nine uranium spheres as a separate matrix region. For simplicity we use a uniform source distribution over each sphere as the source within each region, and find the resulting solution for the system effective multiplication.

The geometry for this problem is shown in Figure 8.6. The nine spheres lie in a rectangular grid with 2 cm of water between them at their closest points. Although it is not stated, we assume the exterior surface of the water is also rectangular in shape. Thus in our geometry there is 2 cm of water between each sphere and the adjacent faces of the system geometry. The outer boundaries of the geometry thus form a parallelepiped that is 7 x 17 x 17 cm. For our calculation each sphere forms one matrix region. The water is not included in the matrix regions but is included in the problem geometry, since it is not necessary to include non-fissioning materials in the matrix regions. This is discussed on p. 243 of the *Primer*.

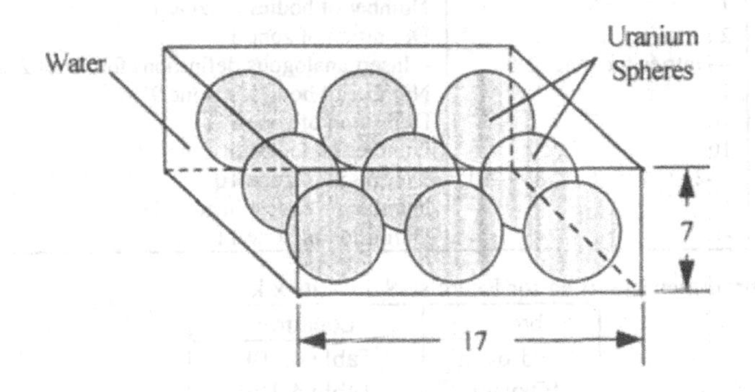

Figure 8.6. Geometry for Exercise 8.3

We define the geometry such that the center of the sphere in the middle of the array is at the origin of the coordinate system and the array of spheres lies in the X-Y plane. The rows of spheres are assumed to lie parallel to these axes. Our resulting geometry definition file 'geom.txt' is shown in Table 8.27. As is clear from this geometry definition, the uranium spheres are defined as zones 1-9, zone 10 is the water surrounding the uranium spheres, and zone 11 is an external void.

The subroutines that are modified for this calculation are listed in Table 8.28. These modifications closely follow those made for Example 8.2, but additional changes are required to accommodate the combination of continuous-energy and multi-group cross sections. The modified subroutine 'Walk' is shown in Table 8.29. This routine is similar to that of Table 8.14P. The number of matrix regions, and the number of iterations to perform on eqns 8.10P and 8.11P, are again stored in common 'in.' In addition, the multi-group cross section variables stored in common 'e82' of Table 8.21 are required for this calculation. In the present problem they are stored in common 'ex83m.' The region number in the loop over regions, 'nreg,' has been added to this common block in order to make the definition of the source region available to other routines.

Table 8.27. Geometry for Exercise 8.3

1	Number of RPPs
-8.5 8.5 -8.5 8.5 -3.5 3.5	Definition of RPP 1, body 1
9	Number of SPHs
-5. -5. 0. 1.5	Definition of SPH 1, body 2
0. -5. 0. 1.5	Definition of SPH 2, body 3
5. -5. 0. 1.5	Definition of SPH 3, body 4
-5. 0. 0. 1.5	Definition of SPH 4, body 5
0. 0. 0. 1.5	Definition of SPH 5, body 6
5. 0. 0. 1.5	Definition of SPH 6, body 7
-5. 5. 0. 1.5	Definition of SPH 7, body 8
0. 5. 0. 1.5	Definition of SPH 8, body 9
5. 5. 0. 1.5	Definition of SPH 9, body 10
11	Number of zones
1	Number of bodies in zone 1
2	Definition of zone 1
– (entries for zones 2-8)	– Insert analogous definitions for zones 2-8
1	Number of bodies in zone 9
10	Definition of zone 9
10	Number of bodies in zone 10
1 -2 -3 -4 -5 -6 -7 -8 -9 -10	Definition of zone 10
1	Number of bodies in zone 11
-1	Definition of zone 11

Table 8.28. Subroutines Modified for Exercise 8.3, Matrix k

Subroutine	Location
'Hydrogen'	Table 4.10P
'Oxygen	Table 4.10P
'Isocol'	Table 6.22
'Xsects'	Table 8.23
'Walk'	Table 8.29
'Input'	Table 8.30
'Source'	Table 8.31
'Col'	Table 8.32
'Stats'	Table 8.33
'keff'	Table 8.34
'Mxsec'	Table 8.35

The modified subroutine 'Input' is shown in Table 8.30. This routine is similar to that of Table 8.15P but the number of regions to be used in the matrix calculation, 'nregions,' is set to nine. The remainder of the problem input is identical to that of Example 8.2.

The modified subroutine 'Source' is shown in Table 8.31. As in Exercise 8.2 the Hansen-Roach multi-group energy bounds and average energies are defined in this routine. The source region is defined by the region variable 'nreg' from subroutine 'Walk' and the value of the source zone, 'nzcur,' is set accordingly. Subroutine 'Source' obtains a source particle location by first selecting a point at random in the middle sphere, zone 5, the center of which is at the origin of the coordinate system. This point is then translated to the correct source sphere by modifying the x and y starting points as appropriate. The z starting point is unchanged by this translation. The energy of the source particle is selected from the Watt fission spectrum, as in Table 4.7P, and the direction of travel is obtained from the library version of subroutine 'Isoout.' Finally, the number of the Hansen-Roach energy group in which the source particle energy lies is determined.

Table 8.29. Subroutine 'Walk' Modified for Exercise 8.3 Matrix k

```
      SUBROUTINE WALK                                          !            1
      COMMON/IN/npart,nregions, iter
      REAL(8) dmfp,dtr,xsec,dcur                               !            3
      COMMON/TRACK/dmfp,dtr,xsec,dcur                          !            4
      REAL(8) x,y,z,u,v,w,xo,yo,zo,uo,vo,wo,wate,age,energ     !            5
      COMMON/PART/x,y,z,u,v,w,xo,yo,zo,uo,vo,wo,wate,age,energ,nzcur,newzn,ngroup  !  6
      COMMON/ex83m/ebound(16),ebar(16),hxs,oxs,uxs,nreg
Loop_Over_Regions:   DO j=1,nregions
  nreg = j
Loop_Over_Particles: DO i=1,npart      ! loop over number of particles              7
      CALL StatLp                      ! initialize statistics for each particle    8
      CALL SOURCE
  Loop_for_Collisions: DO              ! to find collision sites                   10
      CALL DIST                        ! get distance of travel in mfp             11
    Loop_Track:  DO                    ! track particle to collision point         12
      CALL HIT                         ! get distance to boundary of current zone  13
      CALL MXSEC                       ! get total cross section for current zone  14
      IF((dtr-dcur)*xsec.GE.dmfp) EXIT Loop_Track  ! collision before boundary     15
      CALL BDRX                        ! process boundary crossing                 16
      IF(nzcur.EQ.-1) EXIT Loop_for_Collisions  ! particle killed e.g., by escaping 17
    END DO Loop_Track                                          !                   18
      CALL COL                         ! call col to process collision             19
      IF(nzcur.EQ.-1) EXIT Loop_for_Collisions  ! particle killed by collision     20
  END DO Loop_for_Collisions                                   !                   21
      CALL StatELp                     ! process statistics when particle killed   22
END DO Loop_Over_Particles                                     !                   23
END DO Loop_Over_Regions
      RETURN                                                   !                   24
      END                                                      !                   25
```

Table 8.30. Subroutine 'Input' Modified for Exercise 8.3 Matrix k

```
      SUBROUTINE INPUT                                                !       1
      COMMON/IN/npart,nregions,iter
      nregions = 9
      WRITE(16,*)'Number of regions',nregions
      WRITE(*,*)' Enter the number of matrix iterations'
      READ(*,*)iter
      WRITE(16,*)' Number of matrix iterations',iter
      WRITE(*,12)                                                     !       3
   12 FORMAT(' Enter the number of particles per region')
      READ(*,*)npart                                                  !       5
      WRITE(16,*)'Number of particles per region = ',npart
      WRITE(*,13)                                                     !       7
   13 FORMAT(' Enter the random number seed, an integer. ')           !       8
      READ(*,*)iseed                                                  !       9
      IF(iseed.LT.1)iseed=1                                           !      10
      IF(iseed.GE.2147483647)iseed=2147483646                         !      11
      WRITE(16,*)' Starting random number seed is ',iseed             !      12
      CALL rndin(iseed)                                               !      13
      RETURN                                                          !      15
      END                                                             !      16
```

Table 8.31. Modified Subroutine 'Source' for Exercise 8.3 Matrix k

```
      SUBROUTINE SOURCE                                                    !   1
      REAL(8) FLTRN,tmp
      COMMON/IN/npart,nregions,iter
      REAL(8) x,y,z,u,v,w,xo,yo,zo,uo,vo,wo,wate,age,energ                 !   2
      COMMON/PART/x,y,z,u,v,w,xo,yo,zo,uo,vo,wo,wate,age,energ,nzcur,newzn,ngroup !   3
      REAL(8) sphcx(20),sphcy(20),sphcz(20),sphr(20),rpxmin(20),rpxmax(20),&
       rpymin(20),rpymax(20),rpzmin(20),rpzmax(20),bdin(40),bdout(40)
      LOGICAL bdtest(40), bdhit(40)
      COMMON/cg/sphcx,sphcy,sphcz,sphr,rpxmin,rpxmax,rpymin,rpymax,rpzmin,rpzmax,&
       bdin,bdout,nsph,nrpp,nbdy,nzones,nbz(40,41),nnext(40,40,40,2),bdtest,bdhit
      REAL(8) sigt(16),c(16),pscat(16),pabs(16),&
       pfis(16),gpnu(16),pggscat(16,16)
      COMMON/GEOM/sigt,c,pscat,pabs,pfis,gpnu,pggscat,vel,chi,nerggp
!  energ is the current neutron energy in eV
!  ebound are the lower energy bounds of the H-R 16-group set
!  ebar are the group average energies
      COMMON/ex83m/ebound(16),ebar(16),hxs,oxs,uxs,nreg
      DATA ebound/3.e6, 1.4e6, 0.9e6, 0.4e6, 0.1e6, 17000., 3000.,&
       550., 100., 30., 10., 3., 1., 0.4, 0.1, 0.0/
      DATA ebar/4.25e+6, 2.07e+6, 1.13e+6, 0.633e+6, 0.235e+6, &
       38.1e+3, 6.79e+3, 1.2e+3, 222., 53.3, 16.7, 5.32, 1.68, &
       0.621, 0.192, 0.025/
!  flat source assumed in nine spheres
!  source is in matrix region nreg
      nzcur = nreg
!  first find source point in center sphere
      wate =1.0d0
      r=sphr(1)*FLTRN()**0.333333333      !  use for flat source
      costh = -1.+2.*fltrn()
      stheta = sqrt(1.-costh**2)
      phi = 6.2831853*fltrn()
      xctr = r*stheta*cos(phi)
      yctr = r*stheta*sin(phi)
      z = r*costh
!  shift source point to proper sphere
      SELECT CASE(nreg)
        CASE(1,4,7)
          xshift = -5.
        CASE(2,5,8)
          xshift = 0.0
        CASE(3,6,9)
          xshift = 5.
      END SELECT
      SELECT CASE(nreg)
        CASE(1,2,3)
          yshift = -5.
        CASE(4,5,6)
          yshift = 0.
        CASE(7,8,9)
          yshift = 5.
      END SELECT
      x = xctr + xshift
      y = yctr + yshift
!   select source energy from Watt fission spectrum
 65   xtest=-dlog(fltrn());ytest=-dlog(fltrn())      !  pick a point
      IF((ytest-xtest-1.)**2.gt.4.*xtest) GOTO 65   !  reject if fails test
      energ = 2.*xtest*1.e6                        !  value selected from fission spectrum
      DO 300 i=1,nerggp      ! find group number for this energy
      IF(energ.gt.ebound(i))GOTO 301
  300 CONTINUE
  301 ngroup=i
      CALL ISOOUT                  !  direction chosen isotropically          7
      RETURN                                                               !   9
      END                                                                  !  10
```

The modified subroutine 'Col' is shown in Table 8.32. This routine is similar to that of Table 8.17P but is changed to accommodate the current matrix region definitions. Thus if the number of the current zone is less than ten the routine knows that the collision has occurred in one of the uranium spheres. In this case the number of fission neutrons produced in the collision, based on the Hansen-Roach cross sections, is stored in the appropriate element of the fission matrix array 'aij.' The square of the score is stored in the array 'aijsq' in order to estimate the variance. Russian roulette is then played on the post-collision particle.

Table 8.32. Subroutine 'Col' Modified for Exercise 8.3, Matrix k

```
      SUBROUTINE COL                                                         !  1
      REAL(8) FLTRN,delta,r,a
      REAL(8) x,y,z,u,v,w,xo,yo,zo,uo,vo,wo,wate,age,energ                   !  3
      COMMON/PART/x,y,z,u,v,w,xo,yo,zo,uo,vo,wo,wate,age,energ,nzcur,newzn,ngroup  ! 4
      REAL(8) dmfp,dtr,xsec,dcur                                             !  5
      COMMON/TRACK/dmfp,dtr,xsec,dcur                                        !  6
      REAL(8) sigt(16),c(16),pscat(16),pabs(16),&
       pfis(16),gpnu(16),pggscat(16,16)
      COMMON/GEOM/sigt,c,pscat,pabs,pfis,gpnu,pggscat,vel,chi,nerggp
      REAL(8) aij(9,9),aijsq(9,9)
      COMMON/STAT/aij,aijsq
      COMMON/ex83m/ebound(16),ebar(16),hxs,oxs,uxs,nreg
      delta=dmfp/xsec                   ! distance traveled to collision      9
      dtr=dcur+delta                    ! update total distance traveled      10
      x=xo+u*dtr;  y=yo+v*dtr;  z=zo+w*dtr      ! update position             11
      IF(nzcur.LE.9)THEN            ! collision is in a sphere, hits U235
        IF(FLTRN().LE.pfis(ngroup))THEN       ! if fission
          aij(nreg,nzcur) = aij(nreg,nzcur) + wate*gpnu(ngroup)
          aijsq(nreg,nzcur) = aijsq(nreg,nzcur) + (wate*gpnu(ngroup))**2
        ENDIF
        wate=wate*pscat(ngroup)      !  reduce wate by non-absorption
        IF(wate.LT.0.1d0)THEN           !  russian roulette to kill if wate is small
          IF(wate.LT.FLTRN())THEN       !  killed by RR
            nzcur=-1;  RETURN
        ENDIF
          wate=1.0d0                    !  survived, set weight = 1
        ENDIF
        nold = ngroup              ! find new energy group
        r=FLTRN()
        DO 100 i=1,nerggp
          IF(r.LT.pggscat(ngroup,i))GOTO 101
100     CONTINUE
101     ngroup=i
        IF(ngroup.NE.nold) energ = ebar(ngroup)  !  no change for within-group scatter
        CALL ISOOUT                     !  assumes isotropic scatter in lab system   20
        RETURN
      ENDIF
      r = fltrn()             ! collision is in water
      a= 1.
      IF(r.gt.hxs/(hxs+oxs)) a = 16.
      CALL isocol(u,v,w,a,energ)
      IF(energ.lt.ebar(16)) energ = ebar(16)
      DO 4 i = 1,nerggp
        ngroup = i
        IF(energ.gt.ebar(i)) GOTO 5
4     CONTINUE
5     CONTINUE
      RETURN                                                                 !   21
      END                                                                    !   22
```

If the neutron survives the Russian roulette game following a collision with a ^{235}U nucleus a new group for the post-collision neutron is found based on the multi-group scattering matrix, and a new continuous-energy value is assigned to the post-collision neutron. As in Exercise 8.2, if the new group is not the same as the old group this continuous-energy value is defined as the average energy in the new group. If the scattering leaves the neutron in the same energy group, the incident neutron energy is left unchanged. In a collision with a uranium nucleus the post-collision neutron angular distribution is assumed to be isotropic in the laboratory system.

If the collision occurs in zone 10 then the neutron has struck either a hydrogen or an oxygen nucleus. The type of target struck is determined by apportioning the total cross section between H and O at the current neutron energy, and defining the target mass accordingly. The result of the scattering event is obtained from subroutine 'Isocol.' As in Exercise 8.2 this subroutine is identical to that used in Example 4.3 except for the incorporation of double precision where required to be consistent with PFC. Finally, the post-collision energy is tested and, if it has fallen below 0.025 eV, it is set equal to that energy.

The modified subroutine 'Stats' is shown in Table 8.33. The fission-matrix arrays are included in common 'stat' and are set to zero at entry point 'StatOne.' As in the modified version of 'Stats' used in Example 8.2, shown in Table 8.18P, the 'aij' matrix elements are calculated at 'StatEnd' and written to the output file. The system multiplication is determined by calling subroutine 'keff,' shown in Table 8.34. The variance estimate calculated in the latter subroutine does not include any systematic errors introduced by the assumption of a uniform source in each matrix region, and is based on the assumption that the error estimates for the matrix elements are statistically independent. As discussed on p. 244 of the *Primer* we know that this overestimates the uncertainty in the result.

The modified subroutine 'Mxsec' is shown in Table 8.35. In this example the number densities of uranium, hydrogen, and oxygen are obtained based on the full densities of either metallic uranium or water in the appropriate zones. For ^{235}U at 18.80 g/cm^3 we have an atomic density of 0.04818 atoms/barn-cm, while for water at 1 g/cm^3 we have oxygen and hydrogen atom densities of 0.03346 and 0.06692 atoms/barn-cm, respectively. The total microscopic cross sections are obtained on the basis of the zone number in which a collision occurs, and the macroscopic cross sections are calculated using these number densities. The cross section file 'xsects.txt' is identical to that used in Exercise 8.2 and contains the ^{235}U multi-group cross sections from Table 8.31P. The macroscopic cross sections for hydrogen and oxygen are obtained as in Exercise 8.2.

Table 8.33. Subroutine 'Stats' Modified for Exercise 8.3 Matrix k

```
SUBROUTINE Stats                                                              !   1
      COMMON/STAT/aij,aijsq
      REAL(8) aij(9,9),aijsq(9,9)
      COMMON/IN/npart,nregions,iter
      REAL(8) x,y,z,u,v,w,xo,yo,zo,uo,vo,wo,wate,age,energ                    !   5
      COMMON/PART/x,y,z,u,v,w,xo,yo,zo,uo,vo,wo,wate,age,energ,nzcur,newzn,ngroup  !  6
      REAL(8) tmp,tmp1,tmp2,var,stdev,xkeff                                   !   7
      COMMON/ex83m/ebound(16),ebar(16),hxs,oxs,uxs,nreg
ENTRY StatOne          ! entry point to initialize arrays for complete problem   8
  aij = 0.
  aijsq = 0.
RETURN                                                                        !  10
ENTRY StatLp           ! entry point to initialize arrays for a particle         11
RETURN                                                                        !  13
ENTRY StatELp          ! entry point to store scores for a particle              14
RETURN                                                                        !  19
ENTRY StatEBatch
RETURN
ENTRY StatEnd          ! entry point to calculate and print results              20
      WRITE(16,2)
2     FORMAT(/,' keff estimate from matrix k')
      tmp=DFLOAT(npart)
      DO i = 1, nregions
      DO j = 1, nregions
        var=aijsq(i,j)/tmp-(aij(i,j)/tmp)**2
        aij(i,j) = aij(i,j)/tmp
        aijsq(i,j) = var/tmp
      END DO
      END DO
      WRITE(16,*)' aij matrix'
      DO 22 i = 1,9
        WRITE(16,21)(aij(i,j),j=1,9)
22    CONTINUE
21    FORMAT(1p9e10.3)
      CALL keff(xkeff,var)
      WRITE(16,100)xkeff, dsqrt (var)
100   FORMAT(' k-effective = ',1pe14.5,' + or - ',1pe14.5)
RETURN                                                                        !  28
      END                                                                     !  29
```

Running PFC as modified for the Exercise 8.3 matrix k calculation, with 1.5×10^5 start particles per matrix region, gives the matrix results shown in Table 8.36. The diagonal elements of this matrix give the probability of a fission neutron born in each of the uranium spheres producing a fission in the same sphere. Reflection of escaping neutrons by the surrounding water back into the source sphere should be the dominant contribution to the diagonal elements and thus these elements should be approximately equal for all of the spheres. By examining Table 8.36 we see that this is indeed the case.

Other tests of the consistency of the data in Table 8.36 can be made. For example, the coupling between the corner spheres and those adjacent to the corners should be the same; i.e., we should have $a_{12} = a_{14} = a_{32} = a_{36}$, etc. Once again we see that this is approximately the case. We conclude that this fission matrix is at least internally consistent and, apart from the systematic error introduced by the assumption of a uniform source distribution in each region, is probably reasonably accurate. The result obtained for the effective multiplication of this system using this fission matrix and 50 iterations of eqns 8.10P and 8.11P is 0.34710 ± 0.00077.

Table 8.34. Subroutine 'keff' Modified for Exercise 8.3 Matrix k

```
      SUBROUTINE keff(xk,xke)   ! Iterate matrix for criticality
      COMMON/STAT/aij,aijsq
      REAL(8) aij(9,9),aijsq(9,9)
      COMMON/IN/npart,nregions,iter
      REAL(8) s(20),sprime(20),delta(20),xk,xke
      s=1.0;  delta=1.0
 Loop_Over_Iterations: DO n=1,iter    !  begin iteration
      xk = 0.
      DO 1 i = 1,nregions
        xk = xk + s(i)
    1 CONTINUE
      DO 2 i = 1,nregions
        s(i) = s(i)/xk                  !  normalize s
    2 CONTINUE
!       calculate sprime
      DO 3 i = 1,nregions
        sprime(i) = 0.;  delta(i) = 0.
      DO 3 j = 1,nregions
        delta(i) = delta(i)+aijsq(i,j)*s(j)**2
        sprime(i) = sprime(i)+s(j)*aij(i,j)
    3 CONTINUE
      DO 4 i = 1,nregions
        s(i) = sprime(i)
    4 CONTINUE
  END DO Loop_Over_Iterations
      xk = 0.;  xke = 0.
      DO 5 i = 1,nregions
        xke = xke+delta(i)
        xk = xk + s(i)
    5 CONTINUE
      RETURN
      END
```

Table 8.35. Subroutine 'Mxsec' Modified for Exercise 8.3 Matrix k

```
      SUBROUTINE MXSEC                                                          !    1
      REAL(8) dmfp,dtr,xsec,dcur                                                !    2
      COMMON/TRACK/dmfp,dtr,xsec,dcur                                           !    3
      REAL(8) x,y,z,u,v,w,xo,yo,zo,uo,vo,wo,wate,age,energ                      !    4
      COMMON/PART/x,y,z,u,v,w,xo,yo,zo,uo,vo,wo,wate,age,energ,nzcur,newzn,ngroup  !  5
      REAL(8) sigt(16),c(16),pscat(16),pabs(16),&
       pfis(16),gpnu(16),pggscat(16,16)
      COMMON/GEOM/sigt,c,pscat,pabs,pfis,gpnu,pggscat,vel,chi,nerggp
!  current energy e, HR lower bounds 'ebound'
      COMMON/ex83m/ebound(16),ebar(16),hxs,oxs,uxs,nreg
!  number densities for water and U metal
      DATA hdens/0.06692/, odens/0.03346/, udens/0.04818/
      IF(nzcur.GT.9) THEN    !  collision is in water
        e = energ  !  change to single-precision energy variable
        CALL hydrogen(hxs, e, hdens)   !  get hydrogen cross section
        CALL oxygen(oxs, e, odens)     ! get oxygen cross section
        xsec = hxs + oxs
        RETURN
      ENDIF
!  collision is in uranium-235
      uxs=sigt(ngroup)*udens ! get U235 cross section - cross sections not pre-mixed
      xsec = uxs   !  calculate total cross section
      RETURN                                                                    !    9
      END                                                                       !   10
```

Table 8.36. Fission Matrix for Exercise 8.3

	1	2	3	4	5	6	7	8	9
1	2.227-1	2.671-2	4.885-3	2.626-2	1.589-2	3.517-3	4.217-3	3.744-3	1.172-3
2	2.671-2	2.265-1	2.603-2	1.489-2	3.002-2	1.489-2	3.676-3	5.879-3	3.456-3
3	5.155-3	2.688-2	2.214-1	3.477-3	1.609-2	2.666-2	1.674-3	3.574-3	4.721-3
4	2.548-2	1.551-2	3.986-3	2.275-1	2.919-2	5.633-3	2.560-2	1.578-2	3.357-3
5	1.415-2	2.945-2	1.492-2	2.960-2	2.288-1	2.966-2	1.519-2	2.944-2	1.460-2
6	3.966-3	1.538-2	2.675-2	6.429-3	3.016-2	2.274-1	3.561-3	1.444-2	2.629-2
7	4.792-3	3.984-3	1.314-3	2.669-2	1.547-2	4.098-3	2.218-1	2.763-2	5.078-3
8	3.373-3	5.674-3	3.558-3	1.482-2	2.988-2	1.499-2	2.668-2	2.263-1	2.735-2
9	1.272-3	3.646-3	4.711-3	3.749-3	1.514-2	2.737-2	4.560-3	2.710-2	2.243-1

Solution – Matrix by Generation

As in Exercise 8.1, the volumes of the matrix regions do not determine the accuracy of the final criticality result. Instead the accuracy of the result is limited by, among other things, whether or not an accurate source distribution can be estimated for each matrix region. In the current exercise, selecting each of the nine spheres of uranium as a matrix region may require something other than a uniform source distribution across each region to obtain an accurate result for the system multiplication. We can confirm our simple matrix result by checking it with a matrix-by-generation calculation. To do this we use the same geometry as that used in the matrix k calculation above. The PFC subroutines modified for this calculation are listed in Table 8.37.

Table 8.37. Subroutines Modified for Exercise 8.3, Matrix by Generation

Subroutine	Location
'Hydrogen'	Table 4.10P
'Oxygen'	Table 4.10P
'keff'	Table 8.28P
'Walk'	Table 8.19
'Isocol'	Table 6.22
'Xsects'	Table 8.23
'Input'	Table 8.38
'Source'	Table 8.39
'Col'	Table 8.40
'Mxsec'	Table 8.41
'Stats'	Table 8.42

The modified subroutine 'Input' is shown in Table 8.38. This routine is similar to that used in the matrix k calculation above, shown in Table 8.30. The number of batches is added to common 'in' and must be input by the user to perform the matrix-by-generation calculation.

Table 8.38. Subroutine 'Input' for Exercise 8.3 Matrix by Generation

```
      SUBROUTINE INPUT                                                    !      1
      COMMON/IN/npart,nbatch,nregions,iter
      nregions = 9
      WRITE(*,11)
   11 FORMAT(' Enter the number of matrix iterations. ')
      READ(*,*)iter
      WRITE(16,*)' Number of matrix iterations = ',iter
      WRITE(*,12)                                                         !      3
   12 FORMAT(' Enter the number of particles/batch.')
      READ(*,*)npart                                                      !      5
      WRITE(16,*)'Number of particles/batch = ',npart
      WRITE(*,14)
   14 FORMAT(' Enter the number of batches.')
      READ(*,*)nbatch
      WRITE(16,*)'Number of batches ',nbatch
      WRITE(*,13)                                                         !      7
   13 FORMAT(' Enter the random number seed, an integer. ')              !      8
      READ(*,*)iseed                                                      !      9
      IF(iseed.LT.1)iseed=1                                               !     10
      IF(iseed.GE.2147483647)iseed=2147483646                             !     11
      WRITE(16,*)' Starting random number seed is ',iseed                 !     12
      CALL rndin(iseed)                                                   !     13
      RETURN                                                              !     15
      END                                                                 !     16
```

The modified subroutine 'Source' is shown in Table 8.39. This subroutine follows the general form of the version of the same routine shown in Table 8.21 but is specialized for this exercise. Thus at the main entry point this routine defines the first-generation source by selecting source points randomly over the nine uranium spheres. As in Table 8.31 this is done by first selecting a random point in the center sphere and then translating the point to the appropriate sphere.

For each generation the source points are retrieved from the fission bank and loaded into the appropriate tracking variables at entry point 'Sourc2.' In this problem we define the spheres as zones 1-9 and we must add the definition of 'nzcur' in order for the tracking routines to know the zone in which the source particle is born. This is done in a straightforward manner by testing whether the x and y values of the source particle are inside or outside the ranges of ±3 cm. This is possible because the spheres do not overlap in the X-Y plane. Finally, as before, at entry point 'Sourc3' the start-particle energy is selected from the Watt fission spectrum and the starting direction is selected from an isotropic distribution.

The modified subroutine 'Col' is shown in Table 8.40. As in Table 8.32, because this problem involves heterogeneous zones of uranium metal and water, this subroutine distinguishes between collisions in uranium and collisions in water based on the zone number in which the collision occurs. In the case of a collision in water the atomic species of the target is determined using partial cross sections. Otherwise this routine is identical to that of Table 8.22.

Table 8.39. Subroutine 'Source' for Exercise 8.3, Matrix by Generation

```
      SUBROUTINE SOURCE                                              !    1
      REAL(8) FLTRN,tmp
      COMMON/IN/npart,nbatch,nregions,iter
      REAL(8) x,y,z,u,v,w,xo,yo,zo,uo,vo,wo,wate,age,energ           !    2
      COMMON/PART/x,y,z,u,v,w,xo,yo,zo,uo,vo,wo,wate,age,energ,nzcur,newzn,ngroup  !  3
      REAL(8) sphcx(20),sphcy(20),sphcz(20),sphr(20),rpxmin(20),rpxmax(20),&
      rpymin(20),rpymax(20),rpzmin(20),rpzmax(20),bdin(40),bdout(40)
      LOGICAL bdtest(40), bdhit(40)
      COMMON/cg/sphcx,sphcy,sphcz,sphr,rpxmin,rpxmax,rpymin,rpymax,rpzmin,rpzmax,&
       bdin,bdout,nsph,nrpp,nbdy,nzones,nbz(40,41),nnext(40,40,40,2),bdtest,bdhit
      REAL(8) xbank(200000,2),ybank(200000,2),zbank(200000,2),wbank(200000)
      COMMON/BANK/xbank,ybank,zbank,wbank,nsave
      REAL(8) sigt(16),c(16),pscat(16),pabs(16),&
       pfis(16),gpnu(16),pggscat(16,16)
      COMMON/GEOM/sigt,c,pscat,pabs,pfis,gpnu,pggscat,vel,chi,nerggp
!  energ is the current neutron energy in eV
!  ebound are the lower energy bounds of the H-R 16-group set
!  ebar are the group average energies
      REAL(8) bscore,bsumsq,cscore(2),csumsq(2),aij(20,20),aijsq(20,20),&
       bij(20,20),bijsq(20,20)
      COMMON/STAT/bscore,bsumsq,cscore,csumsq,aij,aijsq,&
       bij,bijsq,nstart(20),nbstart(20)
      COMMON/e83mg/ebound(16),ebar(16),hxs,oxs,uxs,nreg
      DATA ebound/3.e6, 1.4e6, 0.9e6, 0.4e6, 0.1e6, 17000., 3000.,&
       550., 100., 30., 10., 3., 1., 0.4, 0.1, 0.0/
      DATA ebar/4.25e+6, 2.07e+6, 1.13e+6, 0.633e+6, 0.235e+6, &
       38.1e+3, 6.79e+3, 1.2e+3, 222., 53.3, 16.7, 5.32, 1.68, &
       0.621, 0.192, 0.025/
!  flat source assumed in nine spheres
!  first find source point in center sphere
      DO 100 i=1,npart
        wbank(i)=1.0d0
        r=sphr(1)*FLTRN()**0.333333333      !  use for flat source
        costh = -1.+2.*fltrn()
        stheta = sqrt(1.-costh**2)
        phi = 6.2831853*fltrn()
        xctr = r*stheta*cos(phi)
        yctr = r*stheta*sin(phi)
        zbank(i,2) = r*costh
!  determine which sphere source point is in
        nsphsce = INT(fltrn()*9.0)+1; IF(nsphsce.GT.9)nsphsce=9
!  source point is in sphere 'nsphsce'.  Adjust x and y coords (+-5)
 2      CONTINUE
        SELECT CASE(nsphsce)
          CASE(1,4,7)
            xshift = -5.
          CASE(2,5,8)
            xshift = 0.0
          CASE(3,6,9)
            xshift = 5.
        END SELECT
        SELECT CASE(nsphsce)
          CASE(1,2,3)
            yshift = -5.
          CASE(4,5,6)
            yshift = 0.
          CASE(7,8,9)
            yshift = 5.
        END SELECT
        xbank(i,2) = xctr + xshift
        ybank(i,2) = yctr + yshift
  100 CONTINUE
      nsave=npart
      RETURN
```

Table 8.39. con't...

```
      ENTRY SOURC2
      initial=npart;  npart=0
      DO 200 i=1,nsave
  150 CONTINUE
      IF(wbank(i).GE.1.0d0)THEN
        npart=npart+1
        xbank(npart,1)=xbank(i,2); ybank(npart,1)=ybank(i,2); zbank(npart,1)=zbank(i,2)
        wbank(i)=wbank(i)-1.0d0
        GOTO 150
      ELSE
        IF(wbank(i).GE.FLTRN())THEN
          npart=npart+1
          xbank(npart,1)=xbank(i,2);ybank(npart,1)=ybank(i,2);zbank(npart,1)=zbank(i,2)
        ENDIF
      ENDIF
  200 CONTINUE
      IF(npart.GT.initial)npart=initial       ! keeps particle numbers from increasing
      nsave=0
      IF (npart.LE.0) THEN
        write(*,*)' The number of particles decreased to zero.'
        STOP
      ENDIF
      IF(npart.LT.initial)THEN                ! keeps particle numbers from decreasing
        ntemp=initial-npart
        DO i=1,ntemp
          npart=npart+1
          xbank(npart,1)=xbank(i,1);ybank(npart,1)=ybank(i,1);zbank(npart,1)=zbank(i,1)
        ENDDO
      ENDIF
      RETURN
      ENTRY SOURC3(n)
      x=xbank(n,1);  y=ybank(n,1);  z=zbank(n,1)
      wate=1.0d0
!  find source zone
      i = 2
      IF(y.LT.-3.)i = 1
      IF(y.GT.3.) i = 3
      j = 2
      IF(x.LT.-3.)j = 1
      IF(x.GT.3.) j = 3
      nzcur = 3*(i-1)+j
!  select energy from Watt fission spectrum
 65   xtest=-dlog(fltrn());ytest=-dlog(fltrn())      !  pick a point
      IF((ytest-xtest-1.)**2.gt.4.*xtest) GOTO 65    !  reject if fails test
      energ = 2.*xtest*1.e6                          !  value selected from fission spectrum
      DO 300 i=1,nerggp        ! find group number for this energy
      IF(energ.gt.ebound(i))GOTO 301
  300 CONTINUE
  301 ngroup=i
      nreg=nzcur;  nstart(nreg)=nstart(nreg)+1;  nbstart(nreg)=nbstart(nreg)+1
      CALL ISOOUT                    !  direction chosen isotropically              7
      RETURN                                                                    !   9
      END                                                                       !  10
```

Subroutine 'Mxsec,' shown in Table 8.41, is the same as that in Table 8.35 except for the use of common 'e83mg.' Subroutine 'Stats' is shown in Table 8.42. This routine is similar to that of Table 8.15.

Table 8.40. Subroutine 'Col' for Exercise 8.3 Matrix by Generation

```
      SUBROUTINE COL                                                    !        1
      REAL(8) FLTRN,delta,r,a
      REAL(8) x,y,z,u,v,w,xo,yo,zo,uo,vo,wo,wate,age,energ              !        3
      COMMON/PART/x,y,z,u,v,w,xo,yo,zo,uo,vo,wo,wate,age,energ,nzcur,newzn,ngroup  !  4
      REAL(8) dmfp,dtr,xsec,dcur                                        !        5
      COMMON/TRACK/dmfp,dtr,xsec,dcur                                   !        6
      REAL(8) sigt(16),c(16),pscat(16),pabs(16),&
       pfis(16),gpnu(16),pggscat(16,16)
      COMMON/GEOM/sigt,c,pscat,pabs,pfis,gpnu,pggscat,vel,chi,nerggp
      REAL(8) bscore,bsumsq,cscore(2),csumsq(2),aij(20,20),aijsq(20,20),&
       bij(20,20),bijsq(20,20),cpart
      COMMON/STAT/bscore,bsumsq,cscore,csumsq,aij,aijsq,&
       bij,bijsq,nstart(20),nbstart(20),cpart
      COMMON/IN/npart,nbatch,nregions,iter
      REAL(8) radius
      REAL(8) xbank(200000,2),ybank(200000,2),zbank(200000,2),wbank(200000)
      COMMON/BANK/xbank,ybank,zbank,wbank,nsave
      COMMON/e83mg/ebound(16),ebar(16),hxs,oxs,uxs,nreg
      delta=dmfp/xsec                   !  distance traveled to collision        9
      dtr=dcur+delta                    !  update total distance traveled       10
      x=xo+u*dtr;  y=yo+v*dtr;  z=zo+w*dtr  !  update position                  11
      IF(nzcur.LE.9) THEN   ! collision is in a sphere, hits U235
        IF(FLTRN().LE.pfis(ngroup))THEN       !  if fission
          nsave=nsave+1                 !  bank a fission
          xbank(nsave,2)=x              !  x-position for fission
          ybank(nsave,2)=y              !  y-position for fission
          zbank(nsave,2)=z              !  z-position for fission
          wbank(nsave)=wate*gpnu(ngroup)    !  weight of fission particle
          cpart=cpart+wate*gpnu(ngroup)
          aij(nreg,nzcur)=aij(nreg,nzcur)+wbank(nsave)   !  scores all fission neutrons
          aijsq(nreg,nzcur)=aijsq(nreg,nzcur)+wbank(nsave)**2
          bij(nreg,nzcur)=bij(nreg,nzcur)+wbank(nsave)   !  scores all fission neutrons
          bijsq(nreg,nzcur)=bijsq(nreg,nzcur)+wbank(nsave)**2
        ENDIF
        wate=wate*pscat(ngroup)         !  reduce wate by non-absorption
        IF(wate.LT.0.1d0)THEN           !  russian roulette to kill if wate is small
          IF(wate.LT.FLTRN())THEN       !  killed by RR
              nzcur=-1;  RETURN
          ENDIF
          wate=1.0d0                    !  survived, set weight = 1
        ENDIF
        nold = ngroup          !  find new energy group
        r=FLTRN()
        DO 100 i=1,nerggp
          IF(r.LT.pggscat(ngroup,i))GOTO 101
 100    CONTINUE
 101    ngroup=i
        IF(ngroup.NE.nold) energ = ebar(ngroup)  !  no change for within-group scatter
        CALL ISOOUT                     !  assumes isotropic scatter in lab system   20
        RETURN
      ENDIF
      r = fltrn()              !  collision is in water
      a= 1.
      IF(r.gt.hxs/(hxs+oxs)) a = 16.
      CALL isocol(u,v,w,a,energ)
      IF(energ.lt.ebar(16)) energ = ebar(16)
      DO 4 i = 1,nerggp
        ngroup = i
        IF(energ.gt.ebar(i)) GOTO 5
 4    CONTINUE
 5    CONTINUE
      RETURN                                                            !       21
      END                                                               !       22
```

Table 8.41. Subroutine 'Mxsec' for Exercise 8.3 Matrix by Generation

```
      SUBROUTINE MXSEC                                                        !   1
      REAL(8) dmfp,dtr,xsec,dcur                                              !   2
      COMMON/TRACK/dmfp,dtr,xsec,dcur                                         !   3
      REAL(8) x,y,z,u,v,w,xo,yo,zo,uo,vo,wo,wate,age,energ                    !   4
      COMMON/PART/x,y,z,u,v,w,xo,yo,zo,uo,vo,wo,wate,age,energ,nzcur,newzn,ngroup  !  5
      REAL(8) sigt(16),c(16),pscat(16),pabs(16),&
       pfis(16),gpnu(16),pggscat(16,16)
      COMMON/GEOM/sigt,c,pscat,pabs,pfis,gpnu,pggscat,vel,chi,nerggp
!  current energy e, HR lower bounds 'ebound'
      COMMON/e83mg/ebound(16),ebar(16),hxs,oxs,uxs,nreg
!  number densities for water and U metal
      DATA hdens/0.06692/, odens/0.03346/, udens/0.04818/
      IF(nzcur.GT.9) THEN    !  collision is in water
        e = energ  !  change to single-precision energy variable
        CALL hydrogen(hxs, e, hdens)   !  get hydrogen cross section
        CALL oxygen(oxs, e, odens)     !  get oxygen cross section
        xsec = hxs + oxs
        RETURN
      ENDIF
!  collision is in uranium-235
      uxs=sigt(ngroup)*udens ! get U235 cross section - cross sections not pre-mixed
      xsec = uxs   !  calculate total cross section
      RETURN                                                                  !   9
      END                                                                     !  10
```

Table 8.42. Subroutine 'Stats' for Exercise 8.3, Matrix-by-Generation

```
 SUBROUTINE Stats                                                             !   1
      REAL(8) bscore,bsumsq,cscore(2),csumsq(2),aij(20,20),aijsq(20,20),&
       bij(20,20),bijsq(20,20),cpart
      COMMON/STAT/bscore,bsumsq,cscore,csumsq,aij,aijsq,&
       bij,bijsq,nstart(20),nbstart(20),cpart
      COMMON/IN/npart,nbatch,nregions,iter
      REAL(8) xbank(200000,2),ybank(200000,2),zbank(200000,2),wbank(200000)
      COMMON/BANK/xbank,ybank,zbank,wbank,nsave
      REAL(8) tmp,tmp1,tmp2,var,stdev,xkeff
      REAL(8) sphcx(20),sphcy(20),sphcz(20),sphr(20),rpxmin(20),rpxmax(20),&
       rpymin(20),rpymax(20),rpzmin(20),rpzmax(20),bdin(40),bdout(40)
      LOGICAL bdtest(40), bdhit(40)
      COMMON/cg/sphcx,sphcy,sphcz,sphr,rpxmin,rpxmax,rpymin,rpymax,rpzmin,rpzmax,&
       bdin,bdout,nsph,nrpp,nbdy,nzones,nbz(40,41),nnext(40,40,40,2),bdtest,bdhit
      COMMON/e83mg/ebound(16),ebar(16),hxs,oxs,uxs,nreg
 ENTRY StatOne              !  entry point to initialize arrays for complete problem  8
      bscore=0.0d0;  bsumsq=0.0d0;  cscore=0.0d0;  csumsq=0.0d0;  outrad=sphr(1)
      aij=0.0d0;  aijsq=0.0d0;  nstart=0;  bij=0.0d0;  bijsq=0.0d0;  nbstart=0
      WRITE(16,1);
      nbchpr = 0
 1    FORMAT(/,' Batch k')
 RETURN                                                                       !  10
 ENTRY StatLp               !  entry point to initialize arrays for a particle   11
      cpart=0.0d0
 RETURN                                                                       !  13
 ENTRY StatELp              !  entry point to store scores for a particle        14
      cscore(1)=cscore(1)+cpart;  csumsq(1)=csumsq(1)+cpart**2
 RETURN                                                                       !  19
 ENTRY StatEBatch                !  entry point to store scores after each batch
      tmp1=DFLOAT(npart);  tmp2=0.d0
      DO 100 i=1,nsave
      tmp2=tmp2+wbank(i)
  100 CONTINUE
      tmp=tmp2/tmp1;  bscore=bscore+tmp;  bsumsq=bsumsq+tmp**2
      nbchpr = nbchpr+1
      WRITE(*,*)  nbchpr, tmp
      WRITE(16,*) nbchpr, tmp
```

Table 8.42. con't...

```
      DO i=1,nregions
        tmp=DFLOAT(nbstart(i))
        DO j=1,nregions
          var=bijsq(i,j)/tmp-(bij(i,j)/tmp)**2
          bij(i,j)=bij(i,j)/tmp
          bijsq(i,j)=var/tmp
        END DO
      END DO
      CALL keff(bij,bijsq,xkeff,var)
      cscore(2)=cscore(2)+xkeff;  csumsq(2)=csumsq(2)+xkeff**2
      bij=0.0d0;  bijsq=0.0d0;  nbstart=0
RETURN
ENTRY StatEnd                   ! entry point to calculate and print results      20
      tmp=DFLOAT(nbatch)
      var=bsumsq/tmp-(bscore/tmp)**2;  stdev=DSQRT(var)
      WRITE(16,*)'ratio of gen      ',bscore/tmp,stdev/DSQRT(tmp)
      tmp=DFLOAT(nbatch)*DFLOAT(npart)
      var=csumsq(1)/tmp-(cscore(1)/tmp)**2;  stdev=DSQRT(var)
      WRITE(16,*)'total subseq pop ',cscore(1)/tmp,stdev/DSQRT(tmp)
      tmp=DFLOAT(nbatch)
      var=csumsq(2)/tmp-(cscore(2)/tmp)**2;  stdev=DSQRT(var)
      WRITE(16,*)'matrix by gen     ',cscore(2)/tmp,stdev/DSQRT(tmp)
      DO i=1,nregions
        tmp=DFLOAT(nstart(i))
        DO j=1,nregions
          var=aijsq(i,j)/tmp-(aij(i,j)/tmp)**2
          aij(i,j)=aij(i,j)/tmp
          aijsq(i,j)=var/tmp
        END DO
      END DO
      CALL keff(aij,aijsq,xkeff,var)
      WRITE(16,200)xkeff,DSQRT(var)
  200 FORMAT(' standard matrix   k-eff = ',1pe17.8,' +- ',1pe17.8)
      WRITE(16,*)(nstart(i),i=1,nregions)
RETURN                                                                !      28
      END                                                             !      29
```

For the matrix-by-generation calculation we again examine the approach to the equilibrium flux by looking at the results for the batch k values while setting 'ntransient' to zero. The results obtained for the batch k values with 1.5 x 10^5 start particles per batch and 50 matrix iterations are plotted in Figure 8.7. As shown in the figure, the uniform-source starting distribution appears to be surprisingly good in that even for the first generation the calculated multiplication factor appears to be a valid estimate of the system multiplication. This indicates that our previous matrix k results are probably fairly accurate. However, to ensure that we have achieved convergence to the equilibrium flux we run the matrix-by-generation calculation with 'ntransient' set equal to five.

The results obtained by running the matrix-by-generation multiplication calculation for Exercise 8.3 with 1.5 x 10^5 start particles per batch, 50 batches, 50 matrix iterations, and 'ntransient' equal to five, are shown in Table 8.43. All of the new results shown in this table are below, but close to, the result of 0.34710 ± 0.00077 obtained from the matrix k calculation for this geometry assuming a uniform source distribution. The latter is shown in the first row of the table.

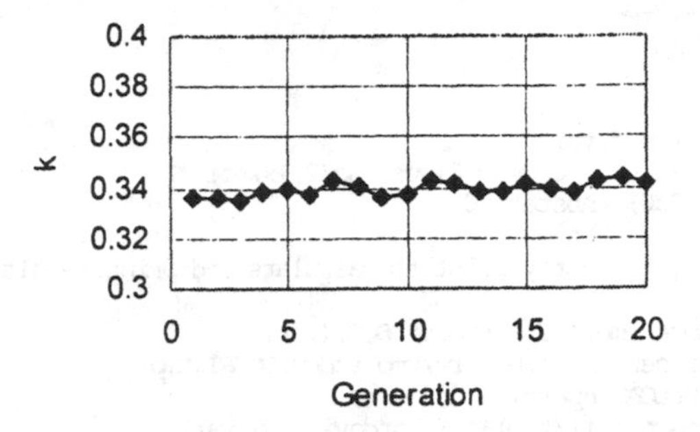

Figure 8.7. Batch k Results for Exercise 8.3 with Uniform Source Distribution in First Generation

Table 8.43. Results for Exercise 8.3

Type of Calculation	Result	Standard Deviation
Standard Matrix (uniform source)	0.347103	0.000769
Ratio of Generations	0.339996	0.000335
Total Subsequent Population	0.339996	0.000332
Matrix by Generation	0.340260	0.000332
Standard Matrix (Generation-based source)	0.340071	0.000320

Chapter 9

Advanced Applications of Monte Carlo

Exercise 9.1

Statement of the Problem

1. Use correlated sampling for the following:
   a. Determine the sensitivity of the particle leakage in Example 6.1 to the total cross section. Assume the value of c does not vary.
   b. Using the results from Example 4.3, determine the sensitivity of the age of thermal neutrons in water from a fission source to the total cross section for neutrons in hydrogen above 1 MeV.

Solution

The sensitivity of the particle leakage in Example 6.1 to the total cross section is determined by using the correlated sampling techniques developed for Example 9.2. Example 6.1 is solved in three different ways to demonstrate several variance reduction techniques. This exercise could be solved without applying variance reduction techniques, but we use source biasing and stratification since this produces the most accurate set of results in the example problem in the *Primer*.

Solution Part a

A list of the PFC routines modified for the present exercise, and the tables in which they are presented, is shown in Table 9.1. The geometry description used for the present problem is that given by Table 6.6P. For the unperturbed solutions we use c = 0.5 and Σ_t = 10.0. For the perturbed solutions the total cross section is changed to Σ_t = 10.1. The results for the change in leakage with cross section, $dL/d\Sigma_t$, obtained by running 10^7 particles with five different seeds for both the unperturbed and perturbed cases, are shown in Table 9.2. The results shown are the average of the right and left leakages since, because of problem symmetry, both are estimates of the same variable. The sum of the two could also have been used as an estimate of total leakage. The average is used here in order to allow easy comparison with the results of Example 6.1.

Table 9.1. Modified Routines for Exercise 9.1a

Routine	Location
'Col'	Table 6.4P
'Bdrx'	Table 6.9P
'Stats'	Table 6.10P
'Source'	Table 6.13P
'Input'	Table 9.7P
'Walk'	Table 9.8P
'sfltrn'	Table 9.2P

Table 9.2. Results for $dL/d\Sigma_t$ for Exercise 9.1a

Seed	Leakage, $\Sigma_t = 10.0$	Leakage, $\Sigma_t = 10.1$	$dL/d\Sigma_t$	$(dL/d\Sigma_t)^2$
1	0.042706	0.042277	-0.00429	1.84041E-5
2	0.042619	0.042196	-0.00423	1.78929E-5
3	0.042681	0.042255	-0.00426	1.81476E-5
4	0.042694	0.042270	-0.00424	1.79776E-5
5	0.042726	0.042297	-0.00429	1.84041E-5

Based on the results shown in Table 9.2, the average value obtained for $dL/d\Sigma_t$ is -0.004262. The average of the sum of the squares is 1.816526×10^{-5}, while the square of the average is 1.8164644×10^{-5}. This gives an estimate for the standard deviation of 1.24×10^{-5}. Our result is thus

$$\left.\frac{dL}{d\Sigma_t}\right|_{\Sigma_t=10} \approx -0.004262 \pm 0.000012$$

Solution Part b

Part b of this exercise calls for the use of the results from Example 4.3 to determine the sensitivity of the age, τ, of thermal neutrons in water from a fission source to the total cross section for neutrons in hydrogen above 1 MeV. We call this quantity $d\tau/d\Sigma_H$. There are two ways to use our previous results to solve this problem. One is to start with the program developed for solving Example 4.3 and add the necessary changes for correlated sampling. The second is to begin with the modified version of PFC presented in Example 7.6, which is based on Example 4.3, and make the required changes to that version of the program. We show both solutions.

The main program and the subroutines needed for the solution based on modifying the program developed for Example 4.3 are listed in Table 9.3. A modified version of the main program used in that example is shown in Table 9.4. Two minor changes have been made in this routine compared with the original version, which is shown in Table 4.7P. First, the definition of the starting random number seed to be used for each source particle now employs the second random number generator 'sfltrn' defined in Example 9.1. Second, the point at which the starting random number is selected for each source particle, performed by a call to subroutine 'rndin,' has been moved inside the 'Loop_Over_Particles.'

Table 9.3. Modified Routines for Exercise 9.1b

Routine	Location
Main	Table 9.4
'Input'	Table 4.8P
'Isocol'	Table 9.5
'Hydrogen'	Table 9.6
'Oxygen'	Table 4.10P
'Sfltrn'	Table 9.2P

Subroutine 'Isocol,' shown in Table 9.5, has been changed from that given in Table 4.9P to eliminate problems caused by rounding errors that occur occasionally when using single-precision variables. Subroutine 'Hydrogen,' shown in Table 9.6, has been modified from that given in Table 4.10P to allow the cross sections for hydrogen above 1.0 MeV to be changed, as required for the perturbation calculation.

The estimates obtained for the value of $d\tau/d\Sigma_H$ by executing these modified routines are shown in Table 9.7 for five different starting random number seeds in the routine 'sfltrn,' using 10^6 start particles per calculation and a cutoff energy of 0.0253 eV. For the perturbed case the cross sections above 1.0 MeV have been increased by 1% over those of the unperturbed case. This is implemented by use of the variable 'delta' in subroutine 'Hydrogen.' For the unperturbed case, 'delta' is set to 1.00. For the perturbed case, 'delta' is set to 1.01 as shown in Table 9.6.

The average value obtained for $d\tau/d\Sigma_H$ is -0.2166. The average of the sum of the squares is 0.046920208. The square of the average is 0.04691556. The resulting estimate of the standard deviation of the average is 1.1×10^{-3}. This gives the rate of change of age per percentage change of hydrogen cross section above 1.0 MeV as $d\tau/d\Sigma_H = -0.2166 \pm 0.0011$.

Table 9.4. Main Program for Exercise 9.1b

```
!                         PROGRAM for Exercise 9.1b Sphere
!    Cross sections for H, O (H in molecular form--not a gas)
!    (cross sections from 0.01 eV to 10 MeV only)
!    Point source of neutrons into uniform sphere, score distance to cutoff energy
!    Scattering is energy-dependent, elastic, isotropic in center of mass.
!    For water (1.0g/cc), set Hdens=0.06692, Odens=0.03346
!    E0 = Source energy, a Watt fission spectrum is used if set=0
!    Ef = cutoff energy;   E = Energy of particle after collision
!    U,V,W = Direction cosines,      X,Y,Z = Cartesian coordinates of particle
!    NMAX = Number of start particles
!    r cm sphere
     double precision fltrn
     Integer sfltrn
     DATA PI/3.14159265/
Loop_Over_Problem: DO
     CALL INPUT(NMAX,E0,Ef,radius,hdens,odens,iseed)
     IF (NMAX.LE.0) STOP                                    !  NMAX=0 to stop
     CALL srndin(iseed)                                      ! set starting random number
     nr=0; nl=0; SCOR=0.; r2bar=0.; r2barsq=0.; N=0         !  initialize for scoring
  Loop_Over_Particles: DO N = 1, NMAX
     Call rndin(iseed)                         ! Modification specific to Exercise 9.1b
     iseed=sfltrn()                            ! Modification specific to Exercise 9.1b
     U = 0.; V = 0.; W = 1.                              !  Source direction cosines
     X = 0.; Y = 0.; Z = 0.                              !  Source location
     IF(e0.le.0.) THEN                                   !  If E0=0.0, use Watt fission spectrum
 65     xtest=-dlog(fltrn());ytest=-dlog(fltrn())          !  pick a point
        IF((ytest-xtest-1.)**2.gt.4.*xtest) GOTO 65       !  reject if fails test
        e = 2.*xtest*1.e6                                !  value selected from fission spectrum
       ELSE                                              !  if not using fission spectrum
        E = E0                                           !  Use input starting energy
     ENDIF
     DO WHILE (e.GT.ef)                                  !  while above cutoff energy
       CALL hydrogen(hxs,e,hdens);CALL oxygen(oxs,e,odens) ! cross sections at energy E
       xs = hxs+oxs                                        !  macro cross section xs in cm**-1
       d = -dlog(fltrn())/xs                               !  Get a flight path D in cm
       X = X+D*U; Y = Y+D*V; Z = Z+D*W                     !  collision site
       r2 = x**2+ y**2+z**2;    r = sqrt(r2)               !  radius of collision site
       IF (r.GE.radius) THEN                             !  if still in sphere, go to 12
         nl = nl + 1                                        !  particle escaped (leaked)
         CYCLE Loop_Over_Particles                          !  get next particle
       ENDIF
       rand = fltrn()                                      !  random # to select type of atom
       IF(rand.lt.hxs/xs)THEN                              !  if collision in hydrogen
        a = 1.                                             !  then A=1
        ELSE                                               !  otherwise collision in oxygen
        a = 16.                                            !  and A=16
       ENDIF
       CALL ISOCOL (u,v,w,A,E)                             !  get new direction and energy
     END DO
     r2bar=r2bar+r2; r2barsq=r2barsq+r2*r2              !  below cutoff energy, score r**2
  END DO Loop_Over_Particles                             !  loop to next particle
     r2bar=r2bar/float(nmax); r2barsq=r2barsq/float(nmax)  !  problem complete
     stdev = sqrt((r2barsq - r2bar*r2bar)/float(nmax))
     WRITE(*,27)r2bar,stdev,r2bar/6.0
 27  FORMAT(' Mean sq dist to abs ',1p2e14.5,' age ',1pe14.5)
     WRITE (*,13)NL
 13  FORMAT(' Particles leaking above ef =',I6)
END DO Loop_Over_Problem
     END
```

Table 9.5. Subroutine 'Isocol' for Exercise 9.1b

```
      SUBROUTINE ISOCOL (U,V,W,A,E)
!     Subroutine ISOCOL calculates the result of elastic collisions
!     of neutrons with nucleii of mass A.  The post-collision
!     direction and energy in the laboratory system is determined
!     assuming isotropic scattering in the center of mass system.
      double precision fltrn
      data twopi/6.2831854/
      COSTH = 2.*fltrn()-1.;       PHI = fltrn()*twopi
      WL = (A*COSTH+1.)/SQRT(A**2+2.*A*COSTH+1.)   !  W in L coordinates
      IF(WL.GE.1.0)WL=.99999999                    ! modified for rounding errors
      SINTH = SQRT(1.-WL**2)
      UL = SINTH*COS(PHI); VL = SINTH*SIN(PHI)     ! U & V in L coords
      E = E*(A*A+2.*A*COSTH+1.)/(A+1.)**2          ! Post-collision energy
      if (abs(u).lt.0.9) then              !  determine which transformation to use
        SR = SQRT(1.-U*U)                  !  x-axis transformation (eqn 4.43)
        Ux = SR*UL + U*WL
        Vx = -U*V*UL/SR + W*VL/SR + V*WL
        Wx = -W*U*UL/SR - V*VL/SR + W*WL
       else
        sr = sqrt(1-v*v)                   !  y-axis transformation (eqn 4.44)
        ux = w*UL/sr - u*v*VL/sr + u*WL
        vx = VL*sr + v*WL
        wx = -u*UL/sr - v*w*VL/sr + w*WL
      endif
      vec = sqrt(ux*ux+vx*vx+wx*wx)        !  normalization to correct for rounding
      u = ux/vec; v = vx/vec; w = wx/vec   !  normalized lab direction cosines
      RETURN;     END
```

Table 9.6. Subroutine 'Hydrogen' for Exercise 9.1b

```
      subroutine HYDROGEN(xs,e,hdens)
!     calculate macroscopic cross section for H,
!     H atom density hdens * 10**24
!     e assumed to be energy in eV
      dimension xspoint(11),epoint(11)
      data epoint/.01,.1,1.,1000.,10000.,50000.,&
      1.e5,1.e6,5.e6,1.e7,3.e7/
      data xspoint/68.,26.,20.,20.,19.,15.,7.,4.,1.5,.9,.3/
      delta=1.01                  ! factor to increase H cross sections over 1 MeV
      do 3 i = 2,11
        ie = i
        if(e.lt.epoint(ie)) go to 4              !  if epoint(ie-1)<energy<epoint(ie)
    3 continue
      xs = hdens*xspoint(ie)*delta       ! Modification specific to Exercise 9.1b
      return
    4 slope = (xspoint(ie)-xspoint(ie-1))/&
      alog(epoint(ie)/epoint(ie-1))
      xs = slope*alog(e/epoint(ie-1))+xspoint(ie-1)    !  extrapolate between ie-1 and ie
      xs = xs*hdens
      IF(ie.ge.9)xs=xs*delta             ! Modification specific to Exercise 9.1b
      return
      end
```

Table 9.7. Results for Exercise 9.1b

Seed	Age, delta = 1.00	Age, delta = 1.01	$d\tau/d\Sigma_H$	$(d\tau/d\Sigma_H)^2$
1	29.2455	29.0257	-0.2198	0.04831204
2	29.2191	29.0036	-0.2155	0.04644025
3	29.1633	28.9500	-0.2133	0.04549689
4	29.1459	28.9290	-0.2169	0.04704561
5	29.1903	28.9728	-0.2175	0.04730625

We now solve this problem using PFC. As the basis for this solution we start with the subroutines used to solve Example 7.6. The listing of the modified subroutines for this exercise is given in Table 9.8. Subroutine 'Bdrx' has been modified to score leaking particles. This routine is shown in Table 9.9. Subroutine 'Col,' shown in Table 9.10, has been modified to perform scattering, as in Example 7.6, by calling 'Isocol,' but to use an energy cut-off rather than a time cut-off.

Table 9.8. Modified Subroutines for Exercise 9.1b with PFC

Subroutine	Location
'Bdrx'	Table 9.9
'Col'	Table 9.10
'Mxsec'	Table 9.11
'Hydrogen'	Table 9.12
'Stats'	Table 9.13
'Xsects'	Table 9.14
'Input'	Table 9.7P
'Isocol'	Table 7.40P
'Oxygen'	Table 7.33P
'sfltrn'	Table 9.2P
'Source'	Table 7.36P
'Walk'	Table 9.8P

Table 9.9. Subroutine 'Bdrx' for Exercise 9.1b with PFC

```
SUBROUTINE BDRX                                                         !   1
REAL(8) delta                                                           !   2
REAL(8) sphcx(20),sphcy(20),sphcz(20),sphr(20),rpxmin(20),rpxmax(20),&  !   3
 rpymin(20),rpymax(20),rpzmin(20),rpzmax(20),bdin(40),bdout(40)         !   4
LOGICAL bdtest(40), bdhit(40)                                           !   5
COMMON/cg/sphcx,sphcy,sphcz,sphr,rpxmin,rpxmax,rpymin,rpymax,rpzmin,rpzmax,&  !  6
 bdin,bdout,nsph,nrpp,nbdy,nzones,nbz(40,41),nnext(40,40,40,2),bdtest,bdhit   !  7
REAL(8) x,y,z,u,v,w,xo,yo,zo,uo,vo,wo,wate,age,energ                    !   8
COMMON/PART/x,y,z,u,v,w,xo,yo,zo,uo,vo,wo,wate,age,energ,nzcur,newzn,ngroup   !  9
REAL(8) dmfp,dtr,xsec,dcur                                              !  10
COMMON/TRACK/dmfp,dtr,xsec,dcur                                         !  11
REAL(8) bscore(10),bsumsq(10),cscore(10),csumsq(10),bpart(10),cpart(10)
COMMON/STAT/bscore,bsumsq,cscore,csumsq,bpart,cpart
delta=dtr-dcur                 ! delta--distance traveled to reach boundary  12
dcur=dtr                       ! update current distance traveled            13
dmfp=dmfp-delta*xsec           ! subtract current distance in mfp from dmfp  14
x=x+u*delta;  y=y+v*delta;  z=z+w*delta    ! update position                 15
nzcur=newzn                    ! change identifier of current zone           16
if(nzcur.LT.nzones)RETURN      ! if not at outer zone then return            17
nzcur=-1                       ! set nzcur=-1 if in outer zone (for escape)  18
bpart(1)=bpart(1)+1
RETURN                                                                  !  19
END                                                                     !  20
```

Subroutine 'Mxsec' is shown in Table 9.11. As in Example 7.6 this routine calls subroutines 'Hydrogen' and 'Oxygen' to obtain the macroscopic total cross sections for these elements. Subroutine 'Hydrogen' is shown in Table 9.12. This routine is similar to that shown in Table 9.6, but includes the modifications shown in Table 7.32P that are associated with the use of double precision variables. Subroutine 'Stats' is shown in Table 9.13. This subroutine calculates the results and writes them on the output file. Subroutine 'Xsects' is shown in Table 9.14. This routine is similar to that in Table 7.34P but the common 'geom' has been modified to be consistent with that in subroutines 'Col' and 'Mxsec.'

Table 9.10. Subroutine 'Col' for Exercise 9.1b with PFC

```
      SUBROUTINE COL                                                         !   1
      REAL(8) FLTRN,delta,A
      REAL(8) x,y,z,u,v,w,xo,yo,zo,uo,vo,wo,wate,age,energ                   !   3
      COMMON/PART/x,y,z,u,v,w,xo,yo,zo,uo,vo,wo,wate,age,energ,nzcur,newzn,ngroup  !  4
      REAL(8) dmfp,dtr,xsec,dcur                                             !   5
      COMMON/TRACK/dmfp,dtr,xsec,dcur                                        !   6
      REAL(8) Hdens,Odens,Hxsec,Oxsec
      COMMON/GEOM/Hdens,Odens,Hxsec,Oxsec ! densities, cross sections
      REAL(8) bscore(10),bsumsq(10),cscore(10),csumsq(10),bpart(10),cpart(10)
      COMMON/STAT/bscore,bsumsq,cscore,csumsq,bpart,cpart
      delta=dmfp/xsec                   !  distance traveled to collision        9
      dtr=dcur+delta                    !  update total distance traveled       10
      x=xo+u*dtr;   y=yo+v*dtr;   z=zo+w*dtr   !  update position               11
      A=16.0d0                          !  scattering might be with oxygen
      IF(fltrn().LT.Hxsec/xsec)A=1.0d0  !  use random number to see if H scattering
      CALL ISOCOL(u,v,w,A,energ)        !  outgoing direction based on atom and energy
      IF(energ.LE.0.0253d0)THEN         !  if below energy cutoff
         cpart(1)=x*x+y*y+z*z
         nzcur=-1
      ENDIF
      RETURN                                                                 !  21
      END                                                                    !  22
```

Table 9.11. Subroutine 'Mxsec' for Exercise 9.1b with PFC

```
      SUBROUTINE MXSEC                                                       !   1
      REAL(8) dmfp,dtr,xsec,dcur                                             !   2
      COMMON/TRACK/dmfp,dtr,xsec,dcur                                        !   3
      REAL(8) x,y,z,u,v,w,xo,yo,zo,uo,vo,wo,wate,age,energ                   !   4
      COMMON/PART/x,y,z,u,v,w,xo,yo,zo,uo,vo,wo,wate,age,energ,nzcur,newzn,ngroup  !  5
      REAL(8) Hdens,Odens,Hxsec,Oxsec
      COMMON/GEOM/Hdens,Odens,Hxsec,Oxsec ! densities, cross sections
      Call Hydrogen(Hxsec,energ,Hdens)       ! returns Hxsec for hydrogen
      Call Oxygen(Oxsec,energ,Odens)         ! Oxsec for oxygen
      xsec=Hxsec+Oxsec
      RETURN                                                                 !   9
      END                                                                    !  10
```

In order to avoid a systematic loss of particles, the geometry of the problem should be sufficiently large that essentially all particles reach the energy cut-off before escaping. The geometry description for this problem is shown in Table 9.15. As is done in Example 4.3, we use a sphere of radius 300 cm. The atomic number densities of H and O for water of nominal density are 0.06692 and 0.03346 atoms/barn-cm, respectively. These values are read from the cross section input file 'xsects.txt' by subroutine 'Xsects,' as shown in Table 9.14, and are used in subroutines 'Hydrogen' and 'Oxygen.'

Table 9.12. Subroutine 'Hydrogen' for Exercise 9.1b with PFC

```
      SUBROUTINE HYDROGEN(xs,e,hdens)
!     calculate macroscopic cross section for H,
!     H atom density hdens * 10**24
!     e assumed to be energy in eV
      DIMENSION xspoint(11),epoint(11)
      REAL(8) xs,e,hdens,delta
      DATA epoint/.01,.1,1.,1000.,10000.,50000.,&
      1.e5,1.e6,5.e6,1.e7,3.e7/
      DATA xspoint/68.,26.,20.,20.,19.,15.,7.,4.,1.5,.9,.3/
      delta=1.01d0
      DO 3 i = 2,11
      ie = i
      if(e.lt.epoint(ie)) GOTO 4                      ! if epoint(ie-1)<energy<epoint(ie)
    3 CONTINUE
      xs = hdens*xspoint(ie)*delta                      ! Energy .GE. 3.E7
      RETURN
    4 slope = (xspoint(ie)-xspoint(ie-1))/&
       ALOG(epoint(ie)/epoint(ie-1))
      xsm = slope*DLOG(e/epoint(ie-1))+xspoint(ie-1)  ! extrapolate between ie-1 and ie
      xs = xsm*hdens
      IF(ie.ge.9)xs=xs*delta
      RETURN
      END
```

Table 9.13. Subroutine 'Stats' for Exercise 9.1b with PFC

```
SUBROUTINE Stats                                                                        !  1
     REAL(8) bscore(10),bsumsq(10),cscore(10),csumsq(10),bpart(10),cpart(10)           !  2
     COMMON/STAT/bscore,bsumsq,cscore,csumsq,bpart,cpart                                !  3
     COMMON/IN/npart,nbatch                                                             !  4
     REAL(8) x,y,z,u,v,w,xo,yo,zo,uo,vo,wo,wate,age,energ                              !  5
     COMMON/PART/x,y,z,u,v,w,xo,yo,zo,uo,vo,wo,wate,age,energ,nzcur,newzn,ngroup    !  6
     REAL(8) tmp,tmp1,tmp2,var,stdev                                                    !  7
ENTRY StatOne              ! entry point to initialize arrays for complete problem  8
     bscore=0.0d0;  bsumsq=0.0d0;  cscore=0.0d0;  csumsq=0.0d0                         !  9
RETURN                                                                                  ! 10
ENTRY StatLp               ! entry point to initialize arrays for a particle          11
     bpart=0.0d0;  cpart=0.0d0                                                          ! 12
RETURN                                                                                  ! 13
ENTRY StatELp              ! entry point to store scores for a particle               14
     bscore(1)=bscore(1)+bpart(1)      ! store score                                  15
     cscore(1)=cscore(1)+cpart(1)      ! store score                                  17
     csumsq(1)=csumsq(1)+cpart(1)**2   ! store square for variance calculation        18
RETURN                                                                                  ! 19
ENTRY StatEnd              ! entry point to calculate and print results               20
     tmp=DFLOAT(npart)                                                                  ! 21
     WRITE(16,*)' number leaked = ', bscore(1)   ! number of particles that leaked
     var=csumsq(1)/tmp-(cscore(1)/tmp)**2        ! variance of cscore distribution  25
     stdev=DSQRT(var)                            ! standard deviation of cscore distr26
     WRITE(16,*)cscore(1)/tmp,stdev/DSQRT(tmp)   ! result and std dev of result      27
     WRITE(16,*)' age = ', cscore(1)/(6.0d0*tmp)
RETURN                                                                                  ! 28
     END                                                                                ! 29
```

The results obtained from running this modified version of PFC for five different starting random number seeds and 10^6 start particles per calculation are given in Table 9.16. The average for $d\tau/d\Sigma_H$ obtained from these results is -0.21592. The average of the sum of the squares is 0.046622832. The square of the average is 0.0466214464. The estimate of the standard deviation of the average is thus 5.9×10^{-4}. This gives the rate of change of age per percentage change of hydrogen cross section above 1.0 MeV as $d\tau/d\Sigma_H$ = -0.21592 ± 0.00059. This compares well with the previous result, -0.2166 ± 0.0011, obtained using the coding based on Example 4.3.

Table 9.14. Subroutine 'Xsects' for Exercise 9.1b with PFC

```
      SUBROUTINE XSECTS                                                    !   1
      REAL(8) Hdens,Odens,Hxsec,Oxsec
      COMMON/GEOM/Hdens,Odens,Hxsec,Oxsec     ! densities, cross sections
      REAL(8) sphcx(20),sphcy(20),sphcz(20),sphr(20),rpxmin(20),rpxmax(20),&   !   4
       rpymin(20),rpymax(20),rpzmin(20),rpzmax(20),bdin(40),bdout(40)      !   5
      LOGICAL bdtest(40), bdhit(40)                                        !   6
      COMMON/cg/sphcx,sphcy,sphcz,sphr,rpxmin,rpxmax,rpymin,rpymax,rpzmin,rpzmax,& !  7
       bdin,bdout,nsph,nrpp,nbdy,nzones,nbz(40,41),nnext(40,40,40,2),bdtest,bdhit  !  8
      OPEN(UNIT=12,FILE='xsects.txt')      ! The cross sections are in xsects.txt   9
!     The outside zone is a vacuum.  particles reaching it have escaped.        10
      DO 50 i=1,nzones-1                                                   !  11
        READ(12,*)Hdens,Odens
        WRITE(16,*)'Densities for Hydrogen and Oxygen in Zone ',i
        WRITE(16,*)Hdens,Odens
   50 CONTINUE                                                             !  15
      RETURN                                                               !  16
      END                                                                  !  17
```

Table 9.15. Geometry Description for Exercise 9.1b with PFC

0	There are no RPPs
1	There is one sphere
0.0d0 0.0d0 0.0d0 300.0d0	Center coordinates and radius-SPH1
2	There are two zones
1 1	Zone 1 has one body, 1
1 -1	Zone 2 has one body, -1

Table 9.16. Results for Exercise 9.1b with PFC

Seed	Age, delta = 1.00	Age, delta = 1.01	$d\tau/d\Sigma_H$	$(d\tau/d\Sigma_H)^2$
1	29.1955	28.9801	-0.2154	0.04639716
2	29.1915	28.9734	-0.2181	0.04756761
3	29.1808	28.9661	-0.2147	0.04609609
4	29.1467	28.9314	-0.2153	0.04635409
5	29.1518	28.9357	-0.2161	0.04669921

Because the current results obtained using PFC are independent of the previous results obtained using the coding from Example 4.3, all ten results can be combined to give an estimate of the rate of change of age per percentage change of hydrogen cross section above 1.0 MeV. This averaged result is $d\tau/d\Sigma_H$ = -0.21626 ± 0.00059.

Exercise 9.2

Statement of the problem

2. Use adjoint transport to determine the probability of a particle passing through a slab two mean free paths thick as a function of the angle at which the particle strikes the slab (see Figure 9.1). Assume isotropic scattering in the laboratory coordinate system with c = 1. [Hint: place an isotropic adjoint source on one face of the slab and score leakage through the opposite face of the slab as a function of angle.] Verify your answer at several angles using forward calculations.

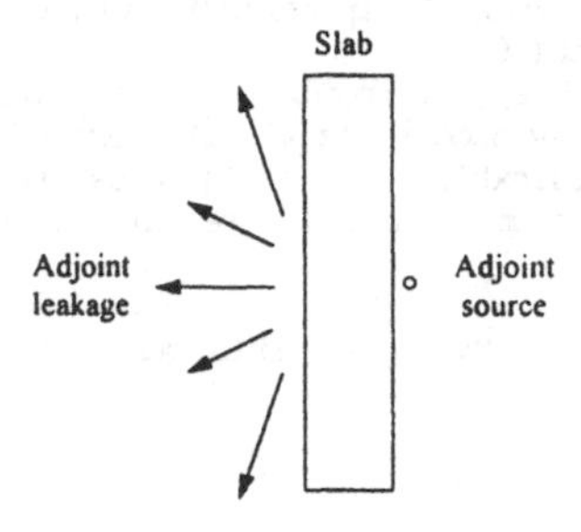

Figure 9.1. Geometry for Exercise 9.2

Solution

In the forward mode we solve this problem by defining a parallel-beam source of particles incident at some particular angle on the upstream face of the slab and we calculate the number of particles transmitted through the slab; i.e., our source is a current of neutrons incident on the slab, J_{in}, and we score the current, J_{out}, that escapes through the downstream face of the slab. The desired solution is the ratio J_{out}/J_{in}. The parallel-beam source could be a pencil beam or it could be uniformly distributed over the surface of the slab with a certain number of incident particles per unit area. To determine the transmission probability as a function of the angle between the incident beam and the normal to the slab we need several such forward calculations, each at a different angle. The angular resolution to which we obtain the answer depends on the number of such calculations performed, and the particular angles chosen for the source incidence. However, instead of running many forward calculations, we can solve this problem with whatever angular resolution we choose by running a single adjoint calculation.

Recall that the vacuum boundary condition for the adjoint flux, which is used in the derivation of eqn 9.31P, is the reverse of that of the forward flux. That is, the outgoing adjoint flux on a surface adjacent to a vacuum is zero,

$$\Psi^{+}(\mathbf{r},\mathbf{\Omega})=0 \text{ for } \mathbf{n}\bullet\mathbf{\Omega}>0 \tag{9.1}$$

In this equation $\mathbf{r}$ is a point on the surface adjacent to the vacuum, and $\mathbf{n}$ is the unit outer normal at this point.

The basic relationship between the forward and adjoint fluxes in a steady-state radiation transport problem is given by eqn 9.32P. Because our problem is monoenergetic we can eliminate the energy integrals in this equation. If we assume both the forward and adjoint sources are constant per unit area over the respective upstream and downstream faces of the slab, both the adjoint and forward fluxes will be independent of position in directions parallel to the faces of the slab. Thus we are left with the forward and adjoint fluxes as functions of the spatial coordinate perpendicular to the slab, which we assume is the Z axis, and of the direction in which the particles are traveling, Ω; i.e., $\Psi = \Psi(z,\Omega)$ and $\Psi^{+} = \Psi^{+}(z,\Omega)$. Eqn 9.32P then becomes

$$\begin{aligned}&\int_{A_o} dA \int_{4\pi} d\mathbf{\Omega}\Psi(z,\mathbf{\Omega})\Psi^{+}(z,\mathbf{\Omega})(\mathbf{n}\bullet\mathbf{\Omega})\\ &\quad = \int_{V_o} dV \int_{4\pi} d\mathbf{\Omega}\left[\Psi^{+}(z,\mathbf{\Omega})S-\Psi(z,\mathbf{\Omega})S^{+}\right]\end{aligned} \tag{9.2}$$

We can write eqn 9.2 for our slab geometry using the variables z and μ, where μ is the cosine of the angle between the particle direction of travel and the positive Z axis. The integral over the area (and over the X and Y coordinates of the volume) can be eliminated because the fluxes do not vary in the X and Y directions. The incident beam source is assumed to be in units of number per unit area. The leakage is also in units of number per unit area. Incorporating these simplifications and applying the vacuum boundary conditions gives

$$\begin{aligned}&-\int_0^1 \Psi_S(0,\mu)\Psi^{+}(0,\mu)\mu d\mu+\int_0^1 \Psi(z_o,\mu)\Psi_S^{+}(z_o,\mu)\mu d\mu\\ &\quad = \int_0^{z_o} dz \int_{-1}^{+1} d\mu\left[\Psi^{+}(z,\mu)S-\Psi(z,\mu)S^{+}\right]\end{aligned} \tag{9.3}$$

Eqn 9.3 takes into account the fact that, in our adjoint formulation, there is no source at $z = 0$ and we have a vacuum boundary condition on the $z = 0$ face of the slab. By eqn 9.1, $\Psi^+(0,\Omega) = 0$ for $\mathbf{n} \bullet \Omega > 0$ and the angular integral of eqn 9.3 need be taken only over incoming values of Ω on this surface. The angular integral on the face at $z = 0$ is negative because for these incoming values of Ω, $\mathbf{n} \bullet \Omega < 0$. At this face $\Psi_S(0,\mu)$ is determined by the forward flux boundary condition on the surface. At the right face of the slab we have no forward source and hence we have a vacuum boundary condition in the forward formulation. Thus the incoming forward angular flux is zero on this surface and we have $\Psi(z_o,\Omega) = 0$ for $\mathbf{n} \bullet \Omega < 0$. As a result the angular integral of eqn 9.3 need be taken only over outgoing values of Ω on the surface $z = z_o$. At this face $\Psi_S^+(z_o,\mu)$ is determined by the adjoint flux boundary condition on the surface.

There are several ways to solve eqn 9.3. Perhaps the most obvious one is to assume that there are no sources inside the volume of the problem for either the forward or adjoint formulation. The forward parallel-beam source is incident on the surface at $z = 0$. The adjoint source is incident on the surface at $z = z_o$. In this case the right side of eqn 9.3 is zero, and the equation becomes

$$\int_0^1 \Psi_S(0,\mu)\Psi^+(0,\mu)\mu d\mu = \int_0^1 \Psi(z_o,\mu)\Psi_S^+(z_o,\mu)\mu d\mu \qquad (9.4)$$

An alternate adjoint formulation can be found by assuming that the surfaces at $z = 0$ and $z = z_o$ are included inside the volume of the problem. There are then sources that are delta functions in the Z coordinate that are internal to the volume, and vacuum boundary conditions exist on the external surfaces for both the forward and adjoint fluxes. In this case both of the terms on the left side of eqn 9.3 are zero, and the integral over the volume on the right side of the equation has a delta function forward source at $z = 0$ and a delta function adjoint source at $z = z_o$. Eqn 9.3 then becomes

$$\int_0^{z_o} dz \int_{-1}^{1} d\mu \Psi^+(z,\mu) S = \int_0^{z_o} dz \int_{-1}^{1} d\mu \Psi(z,\mu) S^+ \qquad (9.5)$$

It is further possible to assume only one of the two sources to be incident on the surface and the other to be a delta function source inside the volume of the problem. This leads to two additional formulations for eqn 9.3,

$$\int_0^1 \Psi_S(0,\mu)\Psi^+(0,\mu)\mu d\mu = \int_0^{z_o} dz \int_{-1}^{1} d\mu \Psi(z,\mu)S^+ \tag{9.6}$$

and

$$\int_0^1 \Psi(z_o,\mu)\Psi_S^+(z_o,\mu)\mu d\mu = \int_0^{z_o} dz \int_{-1}^{1} d\mu \Psi^+(z,\mu)S \tag{9.7}$$

where in eqn 9.6 we assume the adjoint source is inside the volume of the problem, and in eqn 9.7 we assume the forward source is inside the volume of the problem. By comparing eqns 9.6 and 9.7 with eqn 9.5, two other equalities are apparent,

$$\int_0^1 \Psi_S(0,\mu)\Psi^+(0,\mu)\mu d\mu = \int_0^{z_o} dz \int_{-1}^{+1} d\mu \Psi^+(z,\mu)S \tag{9.8}$$

and

$$\int_0^1 \Psi(z_o,\mu)\Psi_S^+(z_o,\mu)\mu d\mu = \int_0^{z_o} dz \int_{-1}^{+1} d\mu \Psi(z,\mu)S^+ \tag{9.9}$$

The fact that the internal delta function forward source and the surface forward source at $z = 0$ are equivalent can be seen from eqn 9.8; i.e.,

$$S = \begin{cases} \delta(z-0)\mu\Psi_S(0,\mu) & \text{for } \mu > 0 \\ 0 & \text{for } \mu < 0 \end{cases} \tag{9.10}$$

Likewise the equivalence of the internal delta function adjoint source and the surface adjoint source at $z = z_o$ can be seen from eqn 9.9; i.e.,

$$S^+ = \begin{cases} \delta(z-z_o)\mu\Psi_S^+(z_o,\mu) & \text{for } \mu > 0 \\ 0 & \text{for } \mu < 0 \end{cases} \tag{9.11}$$

With the above equations it is possible to find explicit expressions for the forward and adjoint sources in this exercise. The forward flux on the surface $z = 0$ is defined by the incoming flux source. This in turn is defined by the incoming current divided by μ ; i.e.,

$$\Psi_S(0,\mu) = \Psi_{inc} = \frac{J_{in}}{\mu}\delta(\mu - \mu_o) \tag{9.12}$$

where J_{in} is the magnitude of the incoming current and μ_o is the cosine of the incident angle. Using eqn 9.12 in eqn 9.10 gives

$$S = \begin{cases} J_{in}\delta(\mu - \mu_o)\delta(z - 0) & \text{for } \mu > 0 \\ 0 & \text{for } \mu < 0 \end{cases} \tag{9.13}$$

To define the adjoint source, recall that this source is related to the response of the detector to the forward flux. For this exercise the response, which we will call R_1, is given by

$$R_1 = J_{out}(z_o) = \int_0^1 \Psi(z_o,\mu)\mu d\mu \tag{9.14}$$

The detector response in terms of the outgoing forward flux at $z = z_o$ is thus a constant times the cosine of the exit angle. For this exercise the constant is one for the forward response. The adjoint source S^+ should then be a delta function at $z = z_o$ and should be a constant times the cosine μ ; i.e.,

$$S^+ = \begin{cases} \mu C\delta(z - z_o) & \text{for } \mu > 0 \\ 0 & \text{for } \mu < 0 \end{cases} \tag{9.15}$$

where C is a constant that represents the strength of the adjoint flux source per unit volume. Using eqn 9.15 in eqn 9.11 gives

$$\Psi_S^{\ +}(z_o,\mu) = C \text{ for } \mu > 0 \tag{9.16}$$

The adjoint source on the surface at $z = z_o$ is therefore an isotropic adjoint flux.

Using the equations for the sources given by eqns 9.12 through 9.16 in any one of the four eqns 9.4 through 9.7 gives the following ratio

$$\frac{J_{out}(z_o)}{J_{in}} = \frac{\Psi^+(0,\mu_o)}{C} \tag{9.17}$$

The left side of eqn 9.17 is the desired solution in the forward formulation; i.e., the probability of leaking through the face at $z = z_o$ per source particle per unit area in the incoming beam. The right side of eqn 9.17 is the adjoint solution that gives the same probability in terms of the adjoint flux at the boundary at $z = 0$ (in the same direction as the incoming forward beam) normalized per unit adjoint flux (isotropic for $\mu > 0$) at the face $z = z_o$. Eqn 9.17 thus relates the ratio of forward currents on the downstream and upstream faces of the slab (the desired result) to the ratio of the adjoint flux on the upstream face and the adjoint source on the downstream face. Solving the adjoint transport equation using the source given by eqn 9.15, and calculating the ratio on the right side of eqn 9.17 then gives the same result as solving the forward problem and calculating the ratio on the left side of eqn 9.17.

As discussed on pages 283-85 of the *Primer*, in order to solve the adjoint transport equation using the Monte Carlo random walk routines developed for solving the forward transport equation a modified adjoint flux Ψ^* is used in which the signs of Ω and Ω' in eqn 9.26P are changed. In this modified formulation adjunctons travel in a direction opposite to forward particles; i.e., they travel from a detector to a source. When using PFC to obtain an adjoint solution to the current exercise, therefore, the modified adjoint source must be placed on the downstream face of the slab with the adjunctons started in the $\mu < 0$ half space. The adjunctons are then scored as they escape through the upstream face. This is shown in Figure 9.1, where the modified adjoint source is on the right, or downstream face of the slab, and the forward source is understood to be on the left, or upstream face of the slab. Applying this change to eqn 9.17 gives

$$\frac{J_{out}(z_o)}{J_{in}} = \frac{\Psi^*(0,-\mu_o)}{C^*} \tag{9.18}$$

The modified adjoint source strength C^* is that of an isotropic flux source of adjunctons on the face $z = z_o$ for $\mu < 0$. It is zero for $\mu > 0$. That is, there are no outgoing adjunctons taken into account during the normalization of the adjoint source strength. To obtain a solution the modified adjoint flux $\Psi^*(z,\mu)$ is scored when adjunctons leak through the face $z = 0$; i.e., the leakage in terms of the true adjoint flux is in the direction of the incoming forward beam.

Monte Carlo transport calculations track particles through the problem geometry and thus directly calculate current, not flux. The source of particles to be tracked in a Monte Carlo transport calculation is always a current, not a flux. For example, in order to estimate the flux produced by particles passing through a surface the particle weights must be divided by the cosine of the angle between the particle direction and the normal to the surface,

$$J = \mu\Psi \text{ or } \Psi = \frac{J}{\mu} \tag{9.19}$$

This relationship between current and flux on a surface is discussed in section 7.4 of the *Primer*. To solve the present problem in the forward mode one would use as a source the quantity J_{in}, which is related to the forward flux by eqn 9.19. Analogously, to solve the problem in the adjoint mode one must use as a source an incoming current associated with the modified adjoint equation. This modified adjoint current is also related to the modified adjoint flux by an equivalent adjoint form of eqn 9.19.

Solution – Adjoint Formulation

Equation 9.18 tells us how to solve the forward transmission problem using the adjoint formulation. It is useful to simplify this equation even further by considering the case of a unit source. For the forward formulation this means that the magnitude of the current J_{in} as used in eqns 9.12 and 9.13 is one, For the adjoint formulation, the magnitude of C is set to one in eqn 9.17, and C* to one in eqn 9.18. As result, eqn 9.18 becomes

$$\tilde{J}_{out}(z_o) = \tilde{\Psi}^*(0,-\mu_o) \tag{9.20}$$

where the symbol ~ indicates the quantity has been normalized and therefore is dimensionless. Here $\tilde{J}_{out}(z_o)$ is the normalized, dimensionless outgoing current for the forward formulation at $z = z_o$ as determined by eqn 9.14.

Now let us consider a second forward problem. For this second problem the desired response is the reflection probability for the forward source as given in eqn 9.13. This response, which we will call R_2, is given by

$$R_2 = J_{out}(0) = \int_0^1 \Psi(0,-\mu)\mu d\mu \tag{9.21}$$

In the original problem the response is given by eqn 9.14 for the leakage out of the slab at $z = z_o$. For this second problem the response is given by eqn 9.21 for the leakage out of the slab at $z = 0$. Assuming this latter leakage is also normalized to a unit source, then $\tilde{J}_{out}(0)$ is the reflection probability for one particle per unit area entering the slab at $z = 0$ with incident angle μ_o, the cosine of the angle at which the forward source is incident on the slab.

Let $\tilde{\Psi}_2^*(z,\mu)$ be the modified adjoint flux solution for this second forward problem. Then

$$\tilde{J}_{out}(0) = \tilde{\Psi}_2^*(0,-\mu_o) \tag{9.22}$$

which is similar to the expression for the transmission probability given by eqn 9.20. The difference is that in this solution for the reflection probability the adjoint source term, in addition to being normalized, is located at the face $z = 0$ instead of $z = z_o$.

Because the slab is symmetric, the solution to a reflection problem with a source at $z = 0$ for which the reflections occur as leakage out of the face $z = 0$ is the same as a reflection problem with a source at $z = z_o$ for which the reflections occur out of the face $z = z_o$. Therefore,

$$\tilde{J}_{out}(0) = \tilde{J}_{out2}(z_o) \tag{9.23}$$

where $\tilde{J}_{out2}(z_o)$ is the current of the reflected particles when the source is incident at $z = z_o$. For this symmetric problem, eqn 9.22 becomes

$$\tilde{J}_{out2}(z_o) = \tilde{\Psi}^*(z_o,\mu_o) \tag{9.24}$$

and the unit adjoint source is at the face $z = z_o$. The normalized adjoint source for the second problem is now at the same face as for the adjoint formulation of the original transmission probability calculation, and the boundary conditions are the same. Therefore, the adjoint Monte Carlo formulations are the same for both problems. The difference is that in the original problem the scoring occurs as adjunctons leak from the face at $z = 0$, whereas in the second problem the scoring occurs as adjunctons leak from the face at $z = z_o$. The same adjoint Monte Carlo random walk can be used to solve for both these reflection and transmission probabilities.

In the forward formulation, because there is no absorption in the slab, the particles must leak and the transmission plus reflection probabilities must add to one. That is

$$\tilde{J}_{out}(z_o) + \tilde{J}_{out}(0) = \tilde{J}_{out}(z_o) + \tilde{J}_{out2}(z_o) = 1 \tag{9.25}$$

and, applying this to the modified adjoint formulation as given in eqns 9.20 and 9.24,

$$\tilde{\Psi}^*(0,-\mu_o) + \tilde{\Psi}^*(z_o,\mu_o) = 1 \quad \text{for } 0 < \mu_o < 1 \tag{9.26}$$

This relationship is true only because of the symmetry of the problem, which allows the equality shown in eqn 9.23. We will make use of this result below.

To use PFC to solve the present adjoint transport exercise, we modify the subroutines listed in Table 9.17. The modified subroutine 'Brdx' is shown in Table 9.18. This routine is used to score the number of adjoint particles that leak from the slab in bins that are defined in terms of the leakage angle; i.e., the cosine of the angle between the direction of the leaking adjuncton and the normal to the slab. Both particles that are transmitted through the slab and particles that are reflected from the slab are scored. The scores, shown in the statement preceding line 19, are divided by the absolute value of the cosine of the exit angle in order to convert the leakage current into a surface flux. Since the cosine bins have a width of 0.1, the scores are divided by 0.1 to normalize to the flux per unit cosine.

Table 9.17. Modified Subroutines for Exercise 9.2, Adjoint Calculation

Subroutine	Location
'Bdrx'	Table 9.18
'Col'	Table 9.19
'Source'	Table 9.20
'Stats'	Table 9.21

Table 9.18. Subroutine 'Bdrx' for Exercise 9.2, Adjoint Calculation

```
SUBROUTINE BDRX                                                        !   1
REAL(8) delta                                                          !   2
REAL(8) sphcx(20),sphcy(20),sphcz(20),sphr(20),rpxmin(20),rpxmax(20),&  !   3
  rpymin(20),rpymax(20),rpzmin(20),rpzmax(20),bdin(40),bdout(40)       !   4
LOGICAL bdtest(40), bdhit(40)                                          !   5
COMMON/cg/sphcx,sphcy,sphcz,sphr,rpxmin,rpxmax,rpymin,rpymax,rpzmin,rpzmax,&  !  6
  bdin,bdout,nsph,nrpp,nbdy,nzones,nbz(40,41),nnext(40,40,40,2),bdtest,bdhit  !  7
REAL(8) x,y,z,u,v,w,xo,yo,zo,uo,vo,wo,wate,age,energ                   !   8
COMMON/PART/x,y,z,u,v,w,xo,yo,zo,uo,vo,wo,wate,age,energ,nzcur,newzn,ngroup  !  9
REAL(8) dmfp,dtr,xsec,dcur                                             !  10
COMMON/TRACK/dmfp,dtr,xsec,dcur                                        !  11
REAL(8) bscore(20),bsumsq(20),cscore(20),csumsq(20),bpart(20),cpart(20)
COMMON/STAT/bscore,bsumsq,cscore,csumsq,bpart,cpart
delta=dtr-dcur                  ! delta--distance traveled to reach boundary  12
dcur=dtr                        ! update current distance traveled            13
dmfp=dmfp-delta*xsec            ! subtract current distance in mfp from dmfp  14
x=x+u*delta;  y=y+v*delta;  z=z+w*delta    ! update position                  15
nzcur=newzn                     ! change identifier of current zone           16
if(nzcur.LT.nzones)RETURN       ! if not at outer zone then return            17
nzcur=-1                        ! set nzcur=-1 if in outer zone (for escape)  18
i=10.0d0*w + 11.0d0;  IF(i.EQ.21)i=20
bpart(i)=wate/DABS(w)/0.1d0
RETURN                                                                 !  19
END                                                                    !  20
```

Subroutine 'Col' is shown in Table 9.19. This subroutine is used to implement an expectation boundary-crossing estimator. Assuming isotropic scattering in the laboratory system, the routine calculates the probability that a post-collision particle will be transmitted through the slab at each of the discrete leakage angles used for scoring. It also calculates the probability that, following the collision, the particle will be reflected from the slab at each of the same leakage angles. The statements between lines number 20 and 21 perform these calculations. The exponential term is equal to the probability that the particle will travel from the collision point to the edge of the slab without a collision, as given by eqn 3.4P. The scores in 'Col' are divided by two since the probability of scattering into a given cosine is normalized by the possible range of the cosine, which varies from minus one to plus one. As in subroutine 'Bdrx' of Table 9.18, the scores in subroutine 'Col' are divided by the absolute value of the cosine of the exit angles in order to convert the exit current into a surface flux.

Table 9.19. Subroutine 'Col' for Exercise 9.2, Adjoint Calculation

```
SUBROUTINE COL                                                        !  1
REAL(8) FLTRN,delta                                                   !  2
REAL(8) x,y,z,u,v,w,xo,yo,zo,uo,vo,wo,wate,age,energ                  !  3
COMMON/PART/x,y,z,u,v,w,xo,yo,zo,uo,vo,wo,wate,age,energ,nzcur,newzn,ngroup  !  4
REAL(8) tmu,d
REAL(8) dmfp,dtr,xsec,dcur                                            !  5
COMMON/TRACK/dmfp,dtr,xsec,dcur                                       !  6
REAL(8) sigt(20),c(20);  !  dimensions allow up to 20 different media    7
COMMON/GEOM/sigt,c; !sigt is total cross section, c is non-absorption prob  8
REAL(8) bscore(20),bsumsq(20),cscore(20),csumsq(20),bpart(20),cpart(20)
COMMON/STAT/bscore,bsumsq,cscore,csumsq,bpart,cpart
delta=dmfp/xsec                 !  distance traveled to collision          9
dtr=dcur+delta                  !  update total distance traveled         10
x=xo+u*dtr;  y=yo+v*dtr;  z=zo+w*dtr    !  update position                11
wate=wate*c(nzcur)              !  reduce wate by non-absorption probability  12
IF(wate.LT.0.1d0)THEN           !  if wate small play russian roulette    13
   IF(wate.LT.FLTRN())THEN      !  if particle killed by RR               14
      nzcur=-1                  !  set nzcur=-1 to show particle killed   15
      RETURN                                                          ! 16
   ENDIF                                                              ! 17
   wate=1.0d0                   !  particle survived RR, increase wate    18
ENDIF                                                                 ! 19
CALL ISOOUT                     !  assumes isotropic scatter in lab system  20
DO i=1,10
  tmu=1.05d0-dfloat(i)*0.1d0; d=(2.0d0-z)/tmu
  cpart(i)=cpart(i)+wate*DEXP(-z/tmu)/(2.0d0*tmu)
  cpart(i+10)=cpart(i+10)+wate*DEXP(-d)/(2.0d0*tmu)
END DO
RETURN                                                                ! 21
END                                                                   ! 22
```

Subroutine 'Source' used for this adjoint calculation is shown in Table 9.20. This routine provides the isotropic adjoint flux source. Only particles entering the slab are used; i.e., all particles start with $w < 0$. The particles are weighted by the cosine of the incident angle to convert the flux on the surface into a current entering the slab.

The modified subroutine 'Stats' is shown in Table 9.21. For each angle bin selected for the calculation, 'Stats' calculates the probability that an adjoint particle will be transmitted and will exit the slab within the given angle interval, as well as the probability that the particle will be reflected and exit within the given angle interval. For the transmission result the

possibility that the source will pass directly through the slab without a collision is included.

Table 9.20. Subroutine 'Source' for Exercise 9.2, Adjoint Calculation

```
      SUBROUTINE SOURCE                                                    !  1
      REAL(8) x,y,z,u,v,w,xo,yo,zo,uo,vo,wo,wate,age,energ                 !  2
      COMMON/PART/x,y,z,u,v,w,xo,yo,zo,uo,vo,wo,wate,age,energ,nzcur,newzn,ngroup  !  3
      pi=2.0d0*DACOS(0.0d0)          ! keep for more complicated source that needs pi  4
      x=0.0d0;  y=0.0d0;  z=2.0d0 ! starts at right edge
      z=z-0.00000001                 ! make sure particle starts inside the slab
      nzcur=1                        ! assumes origin is in zone one         6
      CALL ISOOUT                    ! direction chosen isotropically        7
      w=-DABS(w)                     ! starts in negative z-direction
      wate=DABS(w)                   ! weight to convert to current
      RETURN                                                               !  9
      END                                                                  ! 10
```

Table 9.21. Subroutine 'Stats' for Exercise 9.2, Adjoint Calculation

```
   SUBROUTINE Stats                                                        !  1
      REAL(8) bscore(20),bsumsq(20),cscore(20),csumsq(20),bpart(20),cpart(20)
      COMMON/STAT/bscore,bsumsq,cscore,csumsq,bpart,cpart                  !  3
      COMMON/IN/npart,nbatch                                               !  4
      REAL(8) x,y,z,u,v,w,xo,yo,zo,uo,vo,wo,wate,age,energ                 !  5
      COMMON/PART/x,y,z,u,v,w,xo,yo,zo,uo,vo,wo,wate,age,energ,nzcur,newzn,ngroup  !  6
      REAL(8) tmp,tmp1,tmp2,var,stdev                                      !  7
   ENTRY StatOne          ! entry point to initialize arrays for complete problem  8
      bscore=0.0d0;  bsumsq=0.0d0;  cscore=0.0d0;  csumsq=0.0d0            !  9
   RETURN                                                                  ! 10
   ENTRY StatLp           ! entry point to initialize arrays for a particle   11
      bpart=0.0d0;  cpart=0.0d0                                            ! 12
   RETURN                                                                  ! 13
   ENTRY StatELp          ! entry point to store scores for a particle        14
      DO i=1,20
        bscore(i)=bscore(i)+bpart(i)      ! store score
        bsumsq(i)=bsumsq(i)+bpart(i)**2   ! store square for variance calculation
        cscore(i)=cscore(i)+cpart(i)
        csumsq(i)=csumsq(i)+cpart(i)**2
      END DO
   RETURN                                                                  ! 19
   ENTRY StatEnd          ! entry point to calculate and print results        20
      tmp=DFLOAT(npart)                                                    ! 21
      DO i=1,20
        var=bsumsq(i)/tmp-(bscore(i)/tmp)**2       ! variance of bscore distribution
        stdev=DSQRT(var)                           ! standard deviation of bscore distr
        bscore(i)=bscore(i)/tmp
        WRITE(16,*)bscore(i),stdev/DSQRT(tmp)   ! result and std dev of result
      END DO
      DO i=1,10
        WRITE(16,*)bscore(i)+bscore(21-i)
      END DO
      DO i=1,20
        var=csumsq(i)/tmp-(cscore(i)/tmp)**2       ! variance of cscore distribution
        stdev=DSQRT(var)                           ! standard deviation of cscore distr
        cscore(i)=cscore(i)/tmp
        IF(i.LE.10)cscore(i)=cscore(i)+DEXP(-2.0d0/(1.05d0-DFLOAT(i)*0.1d0))! uncollided
        WRITE(16,*)cscore(i),stdev/DSQRT(tmp)   ! result and std dev of result
      END DO
   RETURN                                                                  ! 28
      END                                                                  ! 29
```

Because the desired ratio for the adjoint solution as given by eqn 9.18 is in terms of the ratio of fluxes, the normalization factor from line 21 in subroutine 'Stats' is the number of particles selected from the isotropic adjoint flux half-space that contributes to the adjoint current entering the slab. Similarly, the values scored when particles leak from the slab are flux values. The ratio of the modified adjoint flux on the face $z = 0$ as a function of the cosine of the leakage angle, normalized by the magnitude of the isotropic flux source strength incident on the face $z = z_o$, then provides the desired ratio given by eqn 9.18.

The 'Bdrx' score provides an average result for the adjoint leakage over the angle intervals (actually over the cosine limits of these angle intervals) specified for the calculation rather than results at specific incident angles. As an angular bin is reduced in size to approximate any specific angle, the statistical uncertainties of the results increase for a given run time. The expectation leakage estimator, on the other hand, estimates the leakage at specific angles and allows the calculation to use the exact cosine of the leakage angle desired.

The geometry description for this exercise is given in Table 9.22. The dimensions of the slab in the X and Y directions have been chosen to simulate an infinite slab. As usual the non-absorption probability and total cross section are defined in the file 'xsects.txt.' By choosing $z_o = 2$, as has been done in the geometry definition, both of these values can be set to one. Results from the adjoint calculation are presented and discussed below.

Table 9.22. Geometry Description for Exercise 9.2

1	There is one RPP
-99.0d0 99.0d0 -99.0d0 99.0d0 0.0d0 2.0d0	xmin, xmax, ymin, ymax, zmin, zmax
0	There are no SPHs
2	There are two zones
1 1	Zone one has one body, 1
1 -1	Zone two has one body, -1

Solution – Forward Formulation

The forward solution to this problem is obtained at a number of different angles in order to compare with the adjoint solution. The modified subroutines for the forward calculation are listed in Table 9.23. Subroutine 'Bdrx' is shown in Table 9.24. The number of particles that are either reflected or transmitted, and thus the current through both faces, is scored in this routine.

The modified subroutine 'Source' for the forward calculation is shown in Table 9.25. The starting angle, contained in the statement following line number 6, can be changed as desired to check the adjoint solution at any desired incident angle. Subroutine 'Stats' is shown in Table 9.26. This routine tallies the scores, calculates the standard deviations, and writes the results to the output file.

Table 9.23. Modified Subroutines for Exercise 9.2, Forward Solution.

Subroutine	Location
'Bdrx'	Table 9.24
'Source'	Table 9.25
'Stats'	Table 9.26

Table 9.24. Subroutine 'Bdrx' for Exercise 9.2, Forward Solution

```
SUBROUTINE BDRX                                                           !  1
REAL(8) delta                                                             !  2
REAL(8) sphcx(20),sphcy(20),sphcz(20),sphr(20),rpxmin(20),rpxmax(20),&    !  3
  rpymin(20),rpymax(20),rpzmin(20),rpzmax(20),bdin(40),bdout(40)          !  4
LOGICAL bdtest(40), bdhit(40)                                             !  5
COMMON/cg/sphcx,sphcy,sphcz,sphr,rpxmin,rpxmax,rpymin,rpymax,rpzmin,rpzmax,& !  6
 bdin,bdout,nsph,nrpp,nbdy,nzones,nbz(40,41),nnext(40,40,40,2),bdtest,bdhit  !  7
REAL(8) x,y,z,u,v,w,xo,yo,zo,uo,vo,wo,wate,age,energ                      !  8
COMMON/PART/x,y,z,u,v,w,xo,yo,zo,uo,vo,wo,wate,age,energ,nzcur,newzn,ngroup !  9
REAL(8) dmfp,dtr,xsec,dcur                                                ! 10
COMMON/TRACK/dmfp,dtr,xsec,dcur                                           ! 11
REAL(8) bscore(2),bsumsq(2),cscore(2),csumsq(2),bpart(2),cpart(2)
COMMON/STAT/bscore,bsumsq,cscore,csumsq,bpart,cpart
delta=dtr-dcur              ! delta--distance traveled to reach boundary  12
dcur=dtr                    ! update current distance traveled            13
dmfp=dmfp-delta*xsec        ! subtract current distance in mfp from dmfp  14
x=x+u*delta; y=y+v*delta;  z=z+w*delta  ! update position                15
nzcur=newzn                 ! change identifier of current zone           16
if(nzcur.LT.nzones)RETURN   ! if not at outer zone then return            17
nzcur=-1                    ! set nzcur=-1 if in outer zone (for escape)  18
i=1;  IF(w.LE.0)i=2
bpart(i)=1.0d0
RETURN                                                                    ! 19
END                                                                       ! 20
```

Table 9.25. Subroutine 'Source' for Exercise 9.2, Forward Solution

```
SUBROUTINE SOURCE                                                         !  1
REAL(8) x,y,z,u,v,w,xo,yo,zo,uo,vo,wo,wate,age,energ                      !  2
COMMON/PART/x,y,z,u,v,w,xo,yo,zo,uo,vo,wo,wate,age,energ,nzcur,newzn,ngroup !  3
REAL(8) pi, phi, sinth
pi=2.0d0*DACOS(0.0d0)          ! keep for more complicated source that needs pi   4
x=0.0d0;  y=0.0d0;  z=0.0d0    ! starts at left face
z=z+0.00000001                 ! to make sure particle starts inside slab
nzcur=1                        ! assumes origin is in zone one                    6
w=0.95d0                       ! starts in positive z-direction
phi=2.0d0*pi*fltrn()
sinth=DSQRT(1.0d0-w**2)
u=DCOS(phi)*sinth;  v=DSIN(phi)*sinth
wate=1.0d0                     ! particle starts with a weight of one             8
RETURN                                                                    !  9
END                                                                       ! 10
```

Results

The results of the adjoint and forward solutions to the current exercise, obtained by running PFC with the modified routines given above, are shown in Table 9.27. All calculations were made using 10^8 source particles. It is obvious that all three sets of results, including those averaged over the bins, are consistent within statistics.

Table 9.26. Subroutine 'Stats' for Exercise 9.2, Forward Solution

```
SUBROUTINE Stats                                                              !  1
     REAL(8) bscore(2),bsumsq(2),cscore(2),csumsq(2),bpart(2),cpart(2)
     COMMON/STAT/bscore,bsumsq,cscore,csumsq,bpart,cpart                      !  3
     COMMON/IN/npart,nbatch                                                   !  4
     REAL(8) x,y,z,u,v,w,xo,yo,zo,uo,vo,wo,wate,age,energ                     !  5
     COMMON/PART/x,y,z,u,v,w,xo,yo,zo,uo,vo,wo,wate,age,energ,nzcur,newzn,ngroup  !  6
     REAL(8) tmp,tmp1,tmp2,var,stdev                                          !  7
ENTRY StatOne                 ! entry point to initialize arrays for complete problem  8
     bscore=0.0d0;  bsumsq=0.0d0;  cscore=0.0d0;  csumsq=0.0d0                !  9
RETURN                                                                        ! 10
ENTRY StatLp                  ! entry point to initialize arrays for a particle   11
     bpart=0.0d0;  cpart=0.0d0                                                ! 12
RETURN                                                                        ! 13
ENTRY StatELp                 ! entry point to store scores for a particle        14
     DO i=1,2
       bscore(i)=bscore(i)+bpart(i)        ! store score
       bsumsq(i)=bsumsq(i)+bpart(i)**2     ! store square for variance calculation
     END DO
RETURN                                                                        ! 19
ENTRY StatEnd                 ! entry point to calculate and print results         20
     tmp=DFLOAT(npart)                                                        ! 21
     DO i=1,2
       var=bsumsq(i)/tmp-(bscore(i)/tmp)**2        ! variance of bscore distribution
       stdev=DSQRT(var)                            ! standard deviation of bscore distr
       WRITE(16,*)bscore(i)/tmp,stdev/DSQRT(tmp)   ! result and std dev of result
     END DO
RETURN                                                                        ! 28
     END                                                                      ! 29
```

The forward calculations required about 12 minutes run time each. A separate forward calculation is needed for each incident angle. The adjoint calculation required about 63 minutes. As discussed above, results for any number of forward incident angles can be obtained with one adjoint calculation, and all of the adjoint results shown in Table 9.27 were obtained in a single calculation. The uncertainties in the adjoint results will increase, for the same run time, as the number of angles selected increases, but this increase will be slow. Because of this relative insensitivity of run time to the number of angles scored, an adjoint calculation is more efficient than a series of forward calculations if several incident angles are to be calculated. If no more than three or four angles are desired, however, then a series of forward calculations would be more efficient. The run time for the adjoint calculation is affected by having scored both the reflected and transmitted adjunctons. This run time could be reduced if only one result were calculated and the other determined by the fact that the transmitted and reflected results should sum to one.

Table 9.27. Results for Exercise 9.2

Incident Cosine (minus for reflected)	Adjoint average over 0.1-width bin	Adjoint using Expectation	Forward
0.95	0.47067 ± 1.9E-4	0.47048 ± 5.6E-5	0.47044 ± 5.0E-5
0.85	0.44466 ± 1.9E-4	0.44502 ± 5.8E-5	0.44494 ± 5.0E-5
0.75	0.41752 ± 1.9E-4	0.41780 ± 5.9E-5	0.41773 ± 4.9E-5
0.65	0.38894 ± 2.0E-4	0.38890 ± 6.1E-5	0.38878 ± 4.9E-5
0.55	0.35844 ± 2.1E-4	0.35852 ± 6.3E-5	0.35842 ± 4.8E-5
0.45	0.32706 ± 2.2E-4	0.32701 ± 6.4E-5	0.32698 ± 4.7E-5
0.35	0.29459 ± 2.4E-4	0.29483 ± 6.6E-5	0.29477 ± 4.6E-5
0.25	0.26211 ± 2.7E-4	0.26223 ± 6.8E-5	0.26217 ± 4.4E-5
0.15	0.22837 ± 3.3E-4	0.22873 ± 7.3E-5	0.22867 ± 4.2E-5
0.05	0.19219 ± 9.4E-4	0.19236 ± 9.7E-5	0.19232 ± 3.9E-5
-0.05	0.80895 ± 1.8E-3	0.80744 ± 1.8E-4	0.80768 ± 3.9E-5
-0.15	0.77102 ± 5.7E-4	0.77127 ± 1.1E-4	0.77133 ± 4.2E-5
-0.25	0.73778 ± 4.3E-4	0.73785 ± 9.6E-5	0.73783 ± 4.4E-5
-0.35	0.70520 ± 3.5E-4	0.70527 ± 8.7E-5	0.70523 ± 4.6E-5
-0.45	0.67296 ± 3.0E-4	0.67310 ± 8.1E-5	0.67302 ± 4.7E-5
-0.55	0.64171 ± 2.7E-4	0.64160 ± 7.6E-5	0.64158 ± 4.8E-5
-0.65	0.61151 ± 2.4E-4	0.61122 ± 7.2E-5	0.61122 ± 4.9E-5
-0.75	0.58252 ± 2.2E-4	0.58232 ± 6.9E-5	0.58227 ± 4.9E-5
-0.85	0.55510 ± 2.0E-4	0.55510 ± 6.6E-5	0.55506 ± 5.0E-5
-0.95	0.52949 ± 1.8E-4	0.52964 ± 6.3E-5	0.52956 ± 5.0E-5

As expected from eqn 9.25 the forward results as shown in the corresponding angle bins for the transmitted and reflected probabilities sum to one. For example, the transmission probability for a cosine of 0.05 is 0.19232. The corresponding reflection probability for a cosine of -0.05 is 0.80768. The sum of these two probabilities is 1.0. Further, the calculated standard deviations for each of these paired probabilities are exactly the same. This is expected from eqn 3.6.

In the results of the adjoint calculation shown in Table 9.27 the transmission and reflection probabilities for a given cosine of the leakage angle sum to one within the statistics of the calculations, as expected from eqn 9.26. In the adjoint case the calculated standard deviations for each of these probabilities are not exactly the same because the results for the reflection and transmission probabilities are obtained from separate estimates. Further, if the slab were not symmetric about $z = 1$, the sums of these two adjoint probabilities would not be one. However, the sum of the two forward probabilities would still be one.

The fact that the reflection and transmission probabilities sum to one within statistics for the adjoint case provides an indication that the sampling of the geometry space of the problem is sufficient to measure effects dependent upon the symmetry in the geometry. The equivalency of the adjoint and forward calculations as well as the match between the boundary-crossing and the expectation boundary-crossing results also provide confidence that the problem geometry has been adequately sampled.

Exercise 9.3

Statement of the problem

3. Example 9.5 assumed all source particles were born with an energy of 0.025 eV and their energy was scored after 20 collisions. Consider the following variations on that calculation.

a. Repeat the example calculation using the technique of tracking a single neutron through many collisions.

b. Use the technique of following many neutrons and scoring the energy of each after some fixed number of collisions, but choose several different start energies. Then, instead of using a fixed start energy, vary the start energies within a single calculation. Did the results change?

c. Using the same parameter values as those used in the example, examine the effect of changing the number of collisions a neutron is allowed to undergo before scoring its energy. Is there an optimum number of collisions that will minimize the run time of the calculation while guaranteeing an equilibrium spectrum?

Solution Part a

The option of tracking a single neutron through many collisions rather than tracking many neutrons through a fixed number of collisions is mentioned in Example 9.5. The program from that example needs to be modified only slightly in order to exercise the former instead of the latter approach. The modified 'Main' routine to be used in the present calculation is shown in Table 9.28. This routine is based on that shown in Table 9.31P. The only significant differences between this routine and that in Table 9.31P are that the initial energy and direction are set outside of the loop over the number of particles, and the call to subroutine 'Therm' is made after each scattering event rather than after 'nscat' scattering events. Other changes that are made include additional write statements to the output file, and the inclusion of a beginning and ending time for the calculation. These additions are made to simplify the comparisons needed for the exercise.

Table 9.28. Main Program for Exercise 9.3a

```
      DIMENSION eg(52),nscor(52)
      DATA bol/8.617342e-5/,pi/3.14159265/
      OPEN(UNIT=16,FILE='output.txt')
      WRITE(*,*)                                        ! Start reading input
  2   WRITE(*,'(1x,a\)')' Enter number of scatters '
      READ(*,*)NMAX;  IF (NMAX.EQ.0) STOP
      WRITE(16,*)' Number of scatters = ',nmax
      WRITE(*,'(1X,A\)')' Enter start random number seed ';  READ(*,*)ISEED
      CALL rndin(iseed)                                 ! set initial random #
      WRITE(16,*)' start random number = ',iseed
      WRITE(*,'(1X,A\)')' enter mass of scatterer '; READ(*,*)a
      WRITE(16,*)' mass of scatterer = ',a
      WRITE(*,'(1X,A\)')' enter temperature (K) ';   READ(*,*)temp  ! End of input
      WRITE(16,*)' temperature (K) = ',temp
      CALL GETTIM(ihr,imin,isec,ihth); WRITE(16,15)ihr,imin,isec ! starting time
 15   FORMAT(     1X,'START TIME: ',I2.2,':',I2.2,':',I2.2)          !
      nSCOR = 0;  ckt=bol*temp; eg(1)=0.0               ! initialize parameters
      DO i= 2,11;  eg(i)=eg(i-1)+0.01;  END DO          ! set energy bounds
      DO i=12,31;  eg(i)=eg(i-1)+0.02;  END DO
      DO i=32,41;  eg(i)=eg(i-1)+0.04;  END DO
      DO i=42,52;  eg(i)=eg(i-1)+0.1 ;  END DO          ! last energy bounds
      e = 0.025;  u = 0.;  v = 0.;  w= 1. ! initial energy and direction, not in loop
      DO N=1,NMAX                            ! loop over NMAX scatters
        CALL Therm(a,e,temp,u,v,w)           ! 'Therm' performs scatter
        Loop_for_index: DO i = 1,52          ! loop to find energy bin
          index=i
          if(e.le.eg(i)) EXIT Loop_for_index     ! if energy bin is found
        END DO Loop_for_index
        nscor(index) = nscor(index)+1            ! score in correct energy bin
      END DO                                     ! End of loop over NMAX particles
      delta=eg(2)-eg(1);  energ=eg(1)+delta/2.0  ! parameters for first energy bin
      sum1=0.0;  sum2=0.0                        ! needed to accumulate scores
      DO 10 i = 2,51                             ! loop over bins, first is empty
        score=FLOAT(nscor(i))/delta/FLOAT(nmax)  ! norm avg score by width of bin
        ac = SQRT(a*energ/bol/temp)
        sigrat = (1.+1./2./ac**2)*erf(ac)+EXP(-ac**2)/ac/1.77245385
        sum1=sum1+score*delta/sigrat;  sum2=sum2+score*delta*energ/sigrat
        delta=eg(i+1)-eg(i);  energ=eg(i)+delta/2.0
 10   CONTINUE
      CALL GETTIM(ihr,imin,isec,ihth); WRITE(16,16)ihr,imin,isec  ! ending time
 16   FORMAT(     1X,'END TIME: ',I2.2,':',I2.2,':',I2.2)
      WRITE(16,*)' Last bin, should be zero', nscor(52)
      WRITE(16,*)' Calc Avg Energy  Theoretical Avg Energy '
      WRITE(16,*)sum2/sum1,2.0*bol*temp
      WRITE(16,*)
      WRITE(16,*)' Energy - Theoretical flux - Calculated flux - Standard Dev '
      delta=eg(2)-eg(1);  energ=eg(1)+delta/2.0
      DO 20 i=2,51
        prob=FLOAT(nscor(i))/FLOAT(nmax)
        StDev=SQRT(prob-prob**2)/SQRT(FLOAT(nmax))    ! binomial distr in eqn 2.29
        score=prob/delta                              ! norm by width of bin
        StDev=StDev/delta
        ac = SQRT(a*energ/bol/temp)
        sigrat = (1.+1./2./ac**2)*erf(ac)+EXP(-ac**2)/ac/1.77245385
        WRITE(16,*)eg(i),energ*EXP(-energ/ckt)/ckt**2,&
          score/sigrat/sum1,StDev/sigrat/sum1
        delta=eg(i+1)-eg(i);  energ=eg(i)+delta/2.0
 20   CONTINUE
      GO TO 2;        END
```

The modified subroutine 'Therm' is shown in Table 9.29. The only changes made to this routine, compared with that of Table 9.32P, are those needed to eliminate two unused variables from 'write' statements (which gave warnings during compilation), and those required to change the value used to adjust for rounding errors from 0.999995 to 0.99995. Because of the use of single precision, without this adjustment fatal rounding errors are encountered on occasion. The function 'Erf,' which is needed in this calculation, is unchanged from that shown in Table 9.32P.

Table 9.29. Subroutine 'Therm' for Exercise 9.3a

```
      SUBROUTINE Therm(a,e,temp,u,v,w)
      REAL(8) fltrn
      DATA bol/8.617342e-5/,pi/3.14159265/          ! bol=Boltzmann const
      alpha = SQRT(a*e/temp/bol)
      sigrat = (1.+1./2./alpha**2)*erf(alpha)+EXP(-alpha**2)/alpha/1.77245385
    1 x1 = 4.*fltrn()                         ! 'x' of eqn 9.48
      x2 = ABS(alpha-x1)**3;  x3 = (alpha+x1)**3-x2
      x4 = 0.45304*x1*x3*EXP(-x1**2)/alpha**2/sigrat
      IF(x4-fltrn().lt.0.) GO TO 1
      x7 = x1/alpha                           ! target velocity
      x5=(alpha+x1)**3
      rand=fltrn()
      xmut=(alpha**2+x1**2-(x5+rand*(x2-x5))**(2./3.))/2./alpha/x1    ! mu of eqn 9.55
      IF(ABS(xmut).lt.1.) GO TO 99
      xmut=xmut*0.99995
      IF(ABS(xmut).lt.1.) GO TO 99
      WRITE(*,*)e,alpha,sigrat,x1,x2,x3,x4,x5,x7,rand,xmut
      STOP 1
   99 b = SQRT(1.-xmut**2)
      rand = fltrn();  c = COS(2.*pi*rand);  d = SIN(2.*pi*rand)
      IF(ABS(w).lt.0.999) GO TO 2
      sqrtu = SQRT(1.-u*u);  wt = b*(c*u*w-d*v)/sqrtu+xmut*w
      vt = b*(c*u*v+d*v)/sqrtu+xmut*v;  ut = xmut*u-b*c*sqrtu
      GO TO 3
    2 sqrtw = SQRT(1.-w*w);  ut = b*(c*w*u-d*v)/sqrtw+xmut*u
      vt = b*(c*w*v+d*u)/sqrtw+xmut*v;  wt = xmut*w-b*c*sqrtw
    3 w0 = 2*fltrn()-1.                       ! get isotropic unit vector
      sinth = SQRT(1.-w0*w0);  phi = 2.*pi*fltrn()
      u0 = sinth*COS(phi);  v0 = sinth*SIN(phi);  y1 = (1.+x7*(x7-2.*xmut))
      IF(y1.gt.0.) GO TO 98
      WRITE(*,*) x7, xmut, y1
      STOP 2
   98 y1 = SQRT(y1);  y3 = u+a*(u0*y1+x7*ut);  y4 = v+a*(v0*y1+x7*vt)
      y5 = w+a*(w0*y1+x7*wt);  y6 = y3**2+y4**2+y5**2
      e = e*y6/(1.+a)**2                      !  outgoing neutron energy and direction
      y2 = 1./SQRT(y6);  u = y3*y2;  v = y4*y2;  w = y5*y2
      RETURN;      END
```

The results obtained from several runs of the above code using A = 235 and T = 300°K are shown in Table 9.30. This heavy target is selected because a high-mass nucleus slows the thermalization process and therefore provides a means of emphasizing the difference between the alternative thermal-spectral calculation methods considered here. In the limit of light targets, hydrogen provides such efficient energy transfer for a neutron that whether one particle is followed for many collisions or many particles are followed for some number of collisions has little effect on the result.

The results shown in Table 9.30 for 10^6 scattering events required less than seven seconds run time. Those for 10^7 scattering events required about

66 seconds run time. Although not shown here, reasonable results for A = 235 and for temperatures of T = 500°K and T = 1000°K were also obtained with 10^7 scattering events and about 66 seconds of run time each. For comparison, a calculation with A = 235, T = 300°K, 'nscat' = 960, and a 0.025-eV starting energy was executed using the coding modifications developed for Example 9.5. With 10^6 start neutrons this calculation provided a thermal distribution with uncertainties comparable to those shown in Table 9.30. However, this latter calculation required about 12 minutes of run time while the results in Table 9.30 were obtained in less than seven seconds. It appears that following one neutron is a much more efficient method for obtaining a thermal neutron spectrum than the method used in Example 9.5 if the complete energy spectrum is desired. On the other hand, from the method demonstrated in the *Primer*, if only a few sample points are needed we have the approximate result that 'ncsat' = 20 + 4A scattering events should suffice for each desired point. Such a calculation would require only a short run time.

Table 9.30. Thermal Neutron Spectra, T = 300°K, One Neutron with Multiple Scatters, A = 235.

Energy (eV)	Theoretical Value	Calculated Value 10^6 Scatters	Calculated Value 10^7 Scatters
1.0E-02	6.1657	5.498 ± 0.023	5.7302 ± 0.0073
2.0E-02	12.564	12.162 ± 0.033	12.392 ± 0.010
3.0E-02	14.222	14.046 ± 0.035	14.196 ± 0.011
4.0E-02	13.524	13.635 ± 0.034	13.555 ± 0.011
5.0E-02	11.810	12.076 ± 0.033	11.851 ± 0.010
6.0E-02	9.8043	10.031 ± 0.030	9.8684 ± 0.0094
7.0E-02	7.8700	8.057 ± 0.027	7.9015 ± 0.0085
8.0E-02	6.1678	6.290 ± 0.024	6.2049 ± 0.0076
9.0E-02	4.7478	4.811 ± 0.021	4.7564 ± 0.0067
1.0E-01	3.6042	3.601 ± 0.019	3.5903 ± 0.0059
0.120	2.3361	2.336 ± 0.011	2.3365 ± 0.0033
0.140	1.2737	1.2452 ± 0.0078	1.2656 ± 0.0025
0.160	0.67798	0.6415 ± 0.0056	0.6617 ± 0.0018
0.180	0.35448	0.3304 ± 0.0041	0.3496 ± 0.0013
0.200	0.18277	0.1641 ± 0.0029	0.1821 ± 9.5-4
0.220	9.3194-02	8.74-2 ± 2.1-3	9.241-2 ± 6.8-4
0.240	4.7088-02	3.98-2 ± 1.4-3	4.376-2 ± 4.7-4
0.260	2.3612-02	2.21-2 ± 1.1-3	2.248-2 ± 3.4-4
Average Energy	0.051704	0.051681	0.051575

Solution Part b

The 'Main' program for this portion of the exercise, as well as that for part c below, is shown in Table 9.31. This program is modified to allow the user to set certain parameters as input to the calculation. Both the particle initial energy and the number of collisions calculated before scoring the particle energy can be defined as input. If the initial energy is set to zero, a random energy from 0.025 to 0.125 eV is selected for each source particle.

Table 9.31. Main Program for Exercise 9.3b and c

```
      DIMENSION eg(52),nscor(52)
      REAL(8) fltrn
      DATA bol/8.617342e-5/,pi/3.14159265/
      OPEN(UNIT=16,FILE='output.txt')
      WRITE(*,*)                                              ! Start reading input
 2    WRITE(*,'(1x,a\)')' Enter number of start particles '
      READ(*,*)NMAX
      IF (NMAX.EQ.0) STOP
      WRITE(16,*)' Number of scatters = ',nmax
      WRITE(*,'(1X,A\)')' Enter start random number seed ';  READ(*,*)ISEED
      CALL rndin(iseed)                                       ! set initial random #
      WRITE(16,*)' start random number = ',iseed
      WRITE(*,'(1X,A\)')' enter mass of scatterer '; READ(*,*)a
      WRITE(16,*)' mass of scatterer = ',a
      WRITE(*,'(1X,A\)')' enter temperature (K) ';   READ(*,*)temp   ! End of input
      WRITE(16,*)' temperature (K) = ',temp
      WRITE(*,'(1X,A\)')' enter No. of fixed scatters '; READ(*,*)nscat
      WRITE(16,*)' No. of fixed scatters = ',nscat
      WRITE(*,'(1X,A\)')' enter start energy, 0 for variable '; READ(*,*)ein
      WRITE(16,*)' start energy= ',ein
      CALL GETTIM(ihr,imin,isec,ihth)  ! gets starting time
      WRITE(16,15)ihr,imin,isec        ! writes starting time
 15   FORMAT(    1X,'START TIME: ',I2.2,':',I2.2,':',I2.2)              !
      nSCOR = 0;  ckt=bol*temp; eg(1)=0.0                      ! initialize parameters
      DO i= 2,11;  eg(i)=eg(i-1)+0.01;  END DO                 ! set energy bounds
      DO i=12,31;  eg(i)=eg(i-1)+0.02;  END DO
      DO i=32,41;  eg(i)=eg(i-1)+0.04;  END DO
      DO i=42,52;  eg(i)=eg(i-1)+0.1 ;  END DO                 ! last energy bounds
      DO N=1,NMAX                                    ! loop over NMAX particles
        e = ein; IF(ein.LE.0.0d0)e=0.025 + fltrn()*.1
        u = 0.;  v = 0.;  w= 1.                      ! initial direction
        DO j = 1,nscat                               ! loop over scatters
          CALL Therm(a,e,temp,u,v,w)                 ! 'Therm' performs scatter
        END DO
        Loop_for_index: DO i = 1,52                  ! loop to find energy bin
          index=i
          if(e.le.eg(i)) EXIT Loop_for_index         ! if energy bin is found
        END DO Loop_for_index
        nscor(index) = nscor(index)+1                ! score in correct energy bin
      END DO                                         ! End of loop over NMAX particles
      delta=eg(2)-eg(1);  energ=eg(1)+delta/2.0      ! parameters for first energy bin
      suml=0.0;  sum2=0.0                            ! needed to accumulate scores
      DO 10 i = 2,51                                 ! loop over bins, first is empty
        score=FLOAT(nscor(i))/delta/FLOAT(nmax)      ! norm avg score by width of bin
        ac = SQRT(a*energ/bol/temp)
        sigrat = (1.+1./2./ac**2)*erf(ac)+EXP(-ac**2)/ac/1.77245385
        suml=suml+score*delta/sigrat;  sum2=sum2+score*delta*energ/sigrat
        delta=eg(i+1)-eg(i);  energ=eg(i)+delta/2.0
 10   CONTINUE
      CALL GETTIM(ihr,imin,isec,ihth)   ! gets ending time
      WRITE(16,16)ihr,imin,isec         ! writes ending time
 16   FORMAT(    1X,'END TIME: ',I2.2,':',I2.2,':',I2.2)
      WRITE(16,*)' Last bin, should be zero', nscor(52)
      WRITE(16,*)' Calc Avg Energy  Theoretical Avg Energy '
      WRITE(16,*)sum2/suml,2.0*bol*temp
      WRITE(16,*)
      WRITE(16,*)' Energy - Theoretical flux - Calculated flux - Standard Dev '
      delta=eg(2)-eg(1);  energ=eg(1)+delta/2.0
```

Table 9.31. con't...

```
      DO 20 i=2,51
        prob=FLOAT(nscor(i))/FLOAT(nmax)
        StDev=SQRT(prob-prob**2)/SQRT(FLOAT(nmax))     ! binomial distr in eqn 2.29
        score=prob/delta                              ! norm by width of bin
        StDev=StDev/delta
        ac = SQRT(a*energ/bol/temp)
        sigrat = (1.+1./2./ac**2)*erf(ac)+EXP(-ac**2)/ac/1.77245385
        WRITE(16,*)eg(i),energ*EXP(-energ/ckt)/ckt**2,&
          score/sigrat/suml,StDev/sigrat/suml
        delta=eg(i+1)-eg(i);  energ=eg(i)+delta/2.0
   20 CONTINUE
      GO TO 2
      END
```

For part b of the exercise, subroutine 'Therm' is unchanged from that of part a. This routine is shown in Table 9.29. Function 'Erf' is shown in Table 9.32P.

To test the effect of different starting energies, A = 235 and 'nscat' = 200 are chosen as input parameters. These values ensure that equilibrium spectra are not obtained. Thus, the effect of the initial energy on the approach to equilibrium is clearly shown by determining how closely the calculated spectrum matches the theoretical equilibrium values. The results obtained from running this problem for start particle energies of 0.025, 0.0517, and random energies between 0.025 and 0.125 eV, all with 10^8 start particles, are shown in Table 9.32.

Table 9.32. Thermal Neutron Flux Spectra, T=300K, A=235, nscat=200

Energy (eV)	Theoretical Value	Calculated Value 'Init' = 0.025	Calculated Value 'Init' = 0.0517	Calculated Value Variable energy
1.0E-02	6.1657	6.82	5.52	4.85
2.0E-02	12.564	14.2	12.2	10.9
3.0E-02	14.222	15.7	14.1	12.7
4.0E-02	13.524	14.3	13.5	12.9
5.0E-02	11.810	12.2	11.9	11.6
6.0E-02	9.8043	9.66	10.0	9.86
7.0E-02	7.8700	7.32	8.01	8.25
8.0E-02	6.1678	5.54	6.24	6.85
9.0E-02	4.7478	3.98	4.90	5.36
1.0E-01	3.6042	3.08	3.69	4.20
0.120	2.3361	1.90	2.37	2.81
0.140	1.2737	0.906	1.29	1.61
0.160	0.67798	0.440	0.653	0.888
0.180	0.35448	0.224	0.329	0.461
0.200	0.18277	0.115	0.173	0.243
0.220	9.3194-2	3.91-2	8.57-2	0.112
0.240	4.7088-2	2.15-2	3.96-2	6.76-2
0.260	2.3612-2	6.51-3	1.65-2	4.11-2
Average	0.051704	0.04677	0.05180	0.05596

From these, admittedly limited, results, it appears that the best starting energy to use to obtain a thermal neutron spectrum is 0.0517 eV, which equals approximately 2kT, the theoretical value of the average energy of the spectrum at T = 300°K, where k is Boltzmann's constant. Starting neutrons with an energy selected randomly between 0.025 and 0.125 eV, the results of which are shown in the last column of Table 9.32, appears to be better than using 0.025 eV as the start energy. The latter is close to the energy of kT ≈ 0.0258 eV associated with the most probable velocity or flux at T = 300°K.

Solution Part c

To solve part c of this exercise we use the routines from part b. However, the number of collisions is now to be varied. Examples for various numbers of collisions, for T = 300°K, 10^8 start particles, and various values of A, are shown in Tables 9.33 to 9.36. In all the cases shown, the starting energy was chosen to be 0.025eV. Although not shown, changing the temperature over the range of 100-1000°K does not result in the need for additional collisions. In the case of hydrogen as a scattering target, as shown in Table 9.36, only ten collisions are needed to obtain reasonable convergence to the equilibrium spectrum. Based on the results of part b of this exercise, even faster convergence should be expected if the theoretical average energy were used as the starting energy.

Table 9.33. Thermal Neutron Spectra, T=300K, A=235, Various 'nscat'

Energy (eV)	Theoretical Value	'Nscat' 240	'Nscat' 480	'Nscat' 720	'Nscat' 960
1.0E-02	6.1657	6.517	5.892	5.807	5.794
2.0E-02	12.564	13.629	12.523	12.314	12.338
3.0E-02	14.222	15.086	14.230	14.185	14.200
4.0E-02	13.524	14.089	13.532	13.586	13.421
5.0E-02	11.810	11.989	11.847	11.709	11.731
6.0E-02	9.8043	9.722	9.804	9.964	9.778
7.0E-02	7.8700	7.582	7.840	7.871	7.918
8.0E-02	6.1678	5.804	6.178	6.148	6.234
9.0E-02	4.7478	4.335	4.677	4.675	4.830
1.0E-01	3.6042	3.180	3.569	3.608	3.603
0.120	2.3361	1.995	2.327	2.364	2.359
0.140	1.2737	1.021	1.257	1.273	1.305
0.160	0.67798	0.5253	0.6750	0.6908	0.6831
0.180	0.35448	0.2578	0.3425	0.3645	0.3571
0.200	0.18277	0.1183	0.1815	0.1826	0.1883
0.220	9.3194-2	6.094-2	8.641-2	9.711-2	8.821-2
0.240	4.7088-2	2.956-2	4.387-2	4.618-2	5.008-2
0.260	2.3612-2	1.434-2	2.007-2	2.357-2	2.520-2
Average	0.051704	0.04827	0.05134	0.05171	0.05180

Table 9.34. Thermal Neutron Spectra, T=300K, A=56, Various 'nscat'

Energy (eV)	Theoretical Value	'Nscat' 76	'Nscat' 136	'Nscat' 188	'Nscat' 244
1.0E-02	6.1657	6.157	5.844	5.829	5.805
2.0E-02	12.564	12.960	12.454	12.363	12.382
3.0E-02	14.222	14.682	14.196	14.142	14.091
4.0E-02	13.524	13.862	13.479	13.498	13.485
5.0E-02	11.810	11.888	11.788	11.764	11.815
6.0E-02	9.8043	9.749	9.835	9.815	9.831
7.0E-02	7.8700	7.734	7.915	7.876	7.883
8.0E-02	6.1678	5.961	6.158	6.199	6.155
9.0E-02	4.7478	4.561	4.734	4.727	4.756
1.0E-01	3.6042	3.412	3.564	3.626	3.608
0.120	2.3361	2.168	2.332	2.371	2.377
0.140	1.2737	1.148	1.279	1.286	1.289
0.160	0.67798	0.6044	0.6831	0.6881	0.6922
0.180	0.35448	0.3022	0.3566	0.3594	0.3564
0.200	0.18277	0.1466	0.1843	0.1859	0.1945
0.220	9.3194-2	7.388-2	9.018-2	9.360-2	9.285-2
0.240	4.7088-2	3.639-2	4.640-2	4.766-2	4.605-2
0.260	2.3612-2	1.956-2	2.207-2	2.404-2	2.288-2
Average	0.051704	0.04993	0.05157	0.05176	0.05178

Table 9.35. Thermal Neutron Spectra, T=300K, A=12, Various 'nscat'

Energy (eV)	Theoretical Value	'Nscat' 20	'Nscat' 32	'Nscat' 44	'Nscat' 56
1.0E-02	6.1657	6.036	5.854	5.798	5.756
2.0E-02	12.564	12.790	12.395	12.392	12.428
3.0E-02	14.222	14.495	14.163	14.140	14.165
4.0E-02	13.524	13.699	13.550	13.467	13.530
5.0E-02	11.810	11.874	11.820	11.822	11.811
6.0E-02	9.8043	9.819	9.840	9.822	9.792
7.0E-02	7.8700	7.757	7.880	7.857	7.872
8.0E-02	6.1678	6.080	6.139	6.173	6.186
9.0E-02	4.7478	4.590	4.730	4.772	4.745
1.0E-01	3.6042	3.465	3.601	3.597	3.603
0.120	2.3361	2.228	2.344	2.365	2.365
0.140	1.2737	1.200	1.273	1.289	1.289
0.160	0.67798	0.6271	0.6847	0.6895	0.6886
0.180	0.35448	0.3209	0.3475	0.3643	0.3480
0.200	0.18277	0.1588	0.1821	0.1821	0.1787
0.220	9.3194-2	7.942-2	9.241-2	9.540-2	9.198-2
0.240	4.7088-2	4.364-2	4.428-2	4.925-2	4.847-2
0.260	2.3612-2	2.007-2	2.394-0	2.310-2	2.269-2
Average	0.051704	0.05052	0.05157	0.05175	0.05169

Discussion

It appears on the basis of the current calculations that the formula used in the *Primer*, 'nscat' = 20 + 4A (see Table 9.31P), is somewhat conservative. Within the accuracy of the present calculations 'nscat' = 20 + 3A appears to converge to the theoretical thermal spectrum as completely as the value used in the *Primer*, and tentatively one may conclude that 'nscat' = 10 + 3A is probably sufficient to achieve equilibrium. However, without a defined criterion of how one may "guarantee" that an equilibrium thermal spectrum exists, it is not possible to select an "optimum" formula for minimizing the run time required to achieve equilibrium.

In the above calculations a comparison between the calculated and the theoretical average energies provides a useful method for determining that the spectrum is reasonable. This is only true, however, because the neutrons are started with an energy equal to about half of the theoretical average energy. If the neutrons had been started with the theoretical average energy, which was shown earlier to be a more efficient scheme for obtaining the thermal spectrum than using a start energy of kT, then the use of the calculated average as a measure of convergence to the correct spectrum would not have been valid.

Table 9.36. Thermal Neutron Spectra, T=300K, A=1, Various 'nscat'

Energy (eV)	Theoretical Value	'Nscat' 4	'Nscat' 10	'Nscat' 20	'Nscat' 24
1.0E-02	6.1657	5.921	5.675	5.631	5.657
2.0E-02	12.564	12.935	12.449	12.420	12.376
3.0E-02	14.222	14.625	14.172	14.174	14.233
4.0E-02	13.524	13.802	13.535	13.536	13.515
5.0E-02	11.810	11.904	11.801	11.767	11.811
6.0E-02	9.8043	9.764	9.719	9.828	9.769
7.0E-02	7.8700	7.755	7.829	7.857	7.889
8.0E-02	6.1678	6.021	6.211	6.178	6.187
9.0E-02	4.7478	4.600	4.760	4.771	4.756
1.0E-01	3.6042	3.413	3.639	3.630	3.598
0.120	2.3361	2.201	2.376	2.378	2.363
0.140	1.2737	1.190	1.301	1.296	1.312
0.160	0.67798	0.6164	0.6863	0.6852	0.6886
0.180	0.35448	0.3101	0.3645	0.3640	0.3609
0.200	0.18277	0.1572	0.1823	0.1864	0.1874
0.220	9.3194-2	7.941-2	9.667-2	9.482-2	9.644-2
0.240	4.7088-2	3.876-2	4.935-2	5.059-2	5.067-2
0.260	2.3612-2	1.997-2	2.475-2	2.614-2	2.299-2
Average	0.051704	0.05031	0.05183	0.05186	0.05184

Index